DATE DUE

NV 27 '99			
FE 8 00			
JY 1 8'01			

DEMCO 38-296

EARTHWORM ECOLOGY
and
BIOGEOGRAPHY
in
NORTH AMERICA

Edited by
PAUL F. HENDRIX

LEWIS PUBLISHERS

Boca Raton   Ann Arbor   London   Tokyo

Library of Congress Cataloging-in-Publication Data

Earthworm ecology and biogeography in North America / edited by Paul F. Hendrix
    p.   cm.
  Includes bibliographical references and index.
  ISBN 1-56670-053-1
  1. Earthworms—Ecology—North America. 2. Earthworms—North America—Geographical
distribution. I. Hendrix, Paul F.
  QL391.A6E26 1995
  595.1′46—dc20   94-23801
                                                          CIP

Gordon E. Gates

1897–1987

This book is dedicated to the memory of Gordon E. Gates, whose lifetime of work laid the foundations for much of our current understanding of earthworms in North America and the world. His contributions are obvious throughout all of the chapters in this book.

Editor

Paul F. Hendrix received his B.S. (1970) and M.S. (1974) in Zoology from Memphis State University, Memphis, Tennessee, and his Ph.D. (1981) in Zoology (Ecology) from the University of Georgia, Athens. He worked as a research biologist with the Environmental Protection Agency from 1981–1983, and as a post-doctoral associate in the Institute of Ecology, University of Georgia, from 1983–1989. Since 1989, he has been an assistant professor jointly staffed in the Institute of Ecology and the Department of Crop and Soil Sciences at the University of Georgia, with research interests in soil ecology and earthworm ecology in native and agricultural ecosystems.

Dr. Hendrix has been an invited speaker at national and international conferences, and has authored or coauthored 58 publications, including coediting one book. Since 1989, he has served as principal investigator on the NSF-funded Horseshoe Bend agroecosystem project at the University of Georgia, and on seven other NSF and USDA funded projects. He has also served as co-principal investigator on six NSF and USDA funded projects since 1986.

Dr. Hendrix's professional activities include membership in the American Association for the Advancement of Science, Soil Science Society of America, Ecological Society of America, and International Soil Science Society.

Preface

Despite the voluminous literature on earthworms, most of our information comes from no more than a few dozen, principally European species; much less is known about the 5,000 or so other species believed to inhabit Earth. And surprisingly, the North American continent ranks among the regions still in need of intensive study, notwithstanding the efforts of a few dedicated specialists over the past several decades.

This situation prompted the present work. In July of 1993, a group of researchers from Mexico, Canada, the U.S., Australia, France, and Germany gathered in the Southern Appalachian mountains of north Georgia, U.S., to consider the state of our understanding of earthworms in North America. Three days of deliberations, followed by several months of revision and editing produced the series of peer-reviewed contributions that comprise this book.

The authors focused on specific topics representing their areas of expertise. They were asked to include (1) a general review and statement of current understanding, including background and historical perspectives; (2) an assessment of current research problems, recent developments and advances, including controversies, unresolved questions, and gaps in knowledge; and (3) a statement of priorities for future research, including promising directions and critical problems needing attention. The result of these efforts is this state-of-the-art treatise and guide for basic and applied research.

Questions concerning biodiversity of soil fauna have received far too little attention in North America, particularly in the U.S. The first half of the book focuses on systematics, biogeography, and ecology of earthworms across the range of climate, soils, and vegetation that occur in North America, from north temperate to tropical. Chapter 1 deals with exotic Lumbricidae, which are now prevalent in Canada and the U.S., while Chapters 2 and 3 cover the nearctic taxa of the southeastern, central, and southwestern U.S., and the Pacific northwest region of the U.S. and Canada. Chapter 4 then covers the neotropical fauna of Mexico, the Caribbean islands, and parts of Central America. All of these chapters include distribution maps of native and selected exotic earthworms, as well as critical reviews of current taxonomy and suggestions for revisions. These chapters clearly show the need for training of new specialists in earthworm taxonomy and biogeography. Chapter 5 considers another question central to earthworm biodiversity—the interactions between native and introduced species in North America and elsewhere. Although direct evidence for competitive exclusion is not strong, distinct stages in the replacement of native earthworms by exotics are recognized. Furthermore, habitat patch size may be crucial to the survival of native populations.

One generalization can be drawn from the past century of research on earthworms: where they are abundant, earthworms can have substantial effects on the structure and function of soils, and therefore are an important consideration in the management of terrestrial ecosystems. The second half of the book addresses these topics. Chapter 6 covers the influence of earthworms on nutrient cycling processes in soil, while Chapter 7 deals with earthworm effects on soil structure, particularly aggregation, porosity, and water and solute transport. The final two chapters focus on earthworms in managed ecosystems, principally agricultural ones, and consider a number of potentially useful applications of our knowledge of earthworm ecology, both in North America and beyond.

The contributions of a number of people are gratefully acknowledged: D.A. Crossley, Jr. first suggested the idea of a workshop on earthworm ecology; H. Ron Pulliam and David E. Kissel provided financial and logistical support for the workshop; and Janice Sand provided hard work and organizing skills to make the workshop successful. Special acknowledgement is given to Nancy Barbour for her masterful work and patience in editing and assembling this book for camera-ready printing. The following people attended the workshop and provided critical comments on the theme papers which have since become the chapters in this book: Isabelle Barois, Mike Beare, Ralph Brinkhurst, George Brown, Mac Callaham Jr., Clarence Callahan, David Coleman, D.A. Crossley, Jr., Bernard Doube, Dory Franklin, John Johnston, Monika Joschko, Eileen Kladivko, Andre Kretschmar, Sarah Lee, Renate Snider, Richard Snider, Bill Tollner, Petra van Vliet, Dan Wallace, Linda Ward, Larry West, and Qiangli Zhang. Their contributions are gratefully acknowledged. The workshop and book preparation were supported by grants from the National Science Foundation and from the U.S. Department of Agriculture, Cooperative State Research Service, National Research Initiative.

Paul F. Hendrix
Athens, Georgia

Contributors

John M. Blair, Division of Biology, Ackert Hall, Kansas State University, Manhattan, Kansas 66506

Patrick J. Bohlen, Institute of Ecosystem Studies, Box AB, Millbrook, New York 12545

Sonia Borges, Departmento de Biologia, RUM, Universidad de Puerto Rico, P.O. Box 5000, Mayaguez, Puerto Rico 00681

Clive A. Edwards, Department of Entomology, 1735 Neil Avenue, Ohio State University, Columbus, Ohio 43210

William M. Edwards, USDA-ARS, P.O. Box 478, Coshocton, Ohio 43812

William M. Fender, Soil Biology Associates, 835 Ashwood Road, McMinnville, Oregon 97128

Carlos Fragoso, Instituto de Ecología, A.C., A.P. 63 C.P. 91000, Xalapa Veracruz, Mexico

Samuel W. James, Department of Biology, Maharishi International University, Fairfield, Iowa 52556

Paul J. Kalisz, Department of Forestry, University of Kentucky, Lexington, Kentucky 40546-0073

Patrick Lavelle, ORSTOM, Centre de Bondy, 72, Route d'Aulnay, 93143 Bondy Cedex, France

Ken E. Lee, CSIRO Division of Soils, Private Bag No. 2, Glen Osmond, S.A. 5064, Australia

Dennis R. Linde, USDA-ARS, Soil Science Building, University of Minnesota, St. Paul, Minnesota 55108

Robert W. Parmelee, Department of Entomology, 1735 Neil Avenue, Ohio State University, Columbus, Ohio 43210

Richard Protz, Land Resource Science, University of Guelph, Guelph, Ontario, Canada N1G 2W1

John W. Reynolds, School of Natural Resources, Sir Sandford Fleming College, P.O. Box 8000, Lindsay, Ontario, Canada, K9V 5E6

Martin J. Shipitalo, USDA-ARS, P.O. Box 478, Coshocton, Ohio 43812

Scott Subler, Department of Entomology, 1735 Neil Avenue, Ohio State University, Columbus, Ohio 43210

Alan P. Tomlin, Agriculture Canada, London Research Center, 1400 Western Road, London, Ontario, Canada N6G 2V4

Hulton B. Wood, USDA-Forest Service, 4955 Canyon Crest Drive, Riverside, California 92507

Contents

Status of Exotic Earthworm Systematics and Biogeography in North America

John W. Reynolds

I.   Introduction

Which species are the exotic earthworms of continental North America, north of Mexico? In the most recent checklist on the subject (Reynolds and Wetzel 1994), in 12 families, excluding the Enchytraeidae, there are 147 species, of which 45 or 31% are introduced. These introduced, or exotic, earthworms are found in primarily two families, i.e., 25 species of Lumbricidae and 14 species of pheretimoid Megascolecidae. This is a very small number, when you consider that the global number of species in each of these groups exceeds 400. This paper will deal with the species in the family Lumbricidae, except for the genera *Bimastos* and *Eisenoides*, which will be covered in one of the presentations on endemic species (Chapter 2).

II.   Background and Historical Perspectives—Systematics

Any discussion of the Lumbricidae must consider the European literature, the centre of origin of this family, and the area exhibiting the greatest species diversity. All historical discussion must begin with Linnaeus, who in his 10th edition of *Systema Naturae* (1758, p. 647) described *Lumbricus terrestris*:

1-56670-053-1/95/$0.00+$.50

LUMBRICUS *Corpus* teres, annulatum. longitudinaliter exasperatum, poro laterali. ...Lumbricus humanus *Habitat* in Humo, adscendit noctu [.] Corpus *annulis ciciter centrum conftans, annuloque majore cartilagineo cinctum, trifariam retrorfum aculeatum; quarto latere inermi...*

In the 10th and subsequent editions of *Systema Naturae*, he still placed all species in the genus *Lumbricus*. There was an amended definition in the 12th edition (1767, p. 1077), which has solidified the opinion that Linnaeus did describe *L. terrestris*, *"exit supra terram tempore nocturno pro copula."*

It was not until 1826, when Savigny was working in Paris, that a second genus was described for the terrestrial earthworms. The mention of the genus *Enterion* is primarily of historical interest. Although most of the 20 or so species described by Savigny still are valid and recognizable today, the genus name has long disappeared, and its species are found in nine of the familiar genera in North America: *Allolobophora*, *Aporrectodea*, *Dendrobaena*, *Dendrodrilus*, *Eisenia*, *Eiseniella*, *Lumbricus*, *Octolasion*, and *Satchellius*.

In 1845, the genus *Helodrilus* was described by Hoffmeister. There has never been a species of this genus recorded from North America.

It was not until 1874 that additional new genera were described. Eisen (1874a, b) added *Allolobophora*, *Allurus* (= *Eiseniella*), and *Dendrobaena*. Many authors over the years have classified the species of *Allolobophora* and *Helodrilus* of Hoffmeister interchangeably.

Malm described the new genus *Eisenia* in 1877.

In 1885, Örley described the new genera *Aporrectodea* and *Octolasion*. In later years, Bouché (1972) and Perel (1976) employed Bouché's synonym *Nicodrilus* for *Aporrectodea*.

The first North American genus to be described in the Lumbricidae was *Bimastos* Moore (1893). Some years earlier (1873–1874), Eisen did describe lumbricid species from initial collections from North America. *Bimastos*, in the strict sense, is a North American genus, and over the years many European authors have placed species of *Bimastos* in *Allolobophora* and occasionally *Helodrilus*.

In 1893, one of the first thorough revisions of the Lumbricidae was made. Rosa divided his species of *Allolobophora* into two subgenera (*Allolobophora* and *Notogama*), plus the genera *Eophila* (described in another paper in 1893), *Dendrobaena*, and *Octolasion*. In the development of taxonomy of earthworms, there have been significant exchanges in attributing the species to one genus or another, and separating and regrouping them into new genera and subgenera. Rosa began a new period, in which the problems of classifying the species into single genera within the Lumbricidae began to be treated from a complex viewpoint.

Beddard (1895), in his monograph, defined the Lumbricidae on five somatic characters and four genital organs. Unfortunately, inexactness then, and on many subsequent occasions, reduced the value of some characters, more particularly the conservative and potentially useful somatic characters.

Michaelsen's *Das Tierreich* (1900) was the first major revision of all the Oligochaeta at the time. In it, he renamed *Allurus* as *Eiseniella*. Since 1900, *Eiseniella* and *Lumbricus* have undergone virtually no changes with respect to the classification of species and subspecies. In Michaelsen's system, he followed that proposed by Rosa in 1893, with only a few minor changes in the status of certain genera and subgenera. Michaelsen classified the genus *Eophila* (Rosa) as a synonym for the genus and subgenus *Helodrilus*. He also classified the genera *Allolobophora* and *Dendrobaena* of Eisen and *Bimastos* of Moore as subgenera of Hoffmeister's *Helodrilus*. Until the revision of Pop (1941), few revisions were made to Michaelsen's classification system.

Smith (1917) relied on the reports and collections of Eisen (1874a, b) and the taxonomy of Michaelsen (1900). In his introduction, Smith surveyed the literature and species described from North American specimens and collections. Smith included only three genera in his Lumbricidae: *Lumbricus*, *Octolasion*, and *Helodrilus*, with its subgenera *Allolobophora*, *Bimastos*, *Dendrobaena*, *Eisenia*, and *Eiseniella*. The spellings, of course, conformed to Michaelsen (1900). It is interesting to note that no new species have been described since Smith's report for Lumbricidae from North America (cf. Species List, *infra*). In his discussion of *Helodrilus (Bimastos)*, he described the species *welchi* which was the last lumbricid species to be described from North America.

Svetlov (1924) and Černosvitov (1935) continued to apply the same lumbricid taxonomic system as established by Rosa and Michaelsen. The taxonomic systems of Rosa and Michaelsen were imperfect and incomplete, based primarily on the number and arrangement of spermathecae. Until the later work by Gates (1972a, c), all lumbricid taxonomic classification systems relied primarily on the highly variable genital characteristics, vis-à-vis the more conservative somatic anatomy. There were some minor differences between Svetlov and Černosvitov and these were in their placing of species in genera and subgenera. Again, practically no changes were made in the species of *Eiseniella* and *Lumbricus*.

Stephenson (1930), in his well known book *The Oligochaeta*, reviewed some of the history of lumbricid taxonomy, but followed the same system as Michaelsen's (1900) *Das Tierreich*.

The first major revision to the classical system of Rosa and Michaelsen came when Pop (1941) pointed out the imperfections of their basic taxonomic characters, i.e. number of seminal vesicles and the position of spermathecae. A summary of Pop's contributions includes:
- eliminating the genera *Bimastos* and *Eophila*,
- modifying the delimitation of the genera *Allolobophora*, *Eisenia* and *Dendrobaena*,
- removing from *Eisenia* species with separate (distinct) setae and placing them into the genus *Dendrobaena*, and species with widely-paired setae into the genus *Allolobophora*,
- removing from *Dendrobaena* all species with closely-paired setae,
- considering body color and setal arrangement, he classified the species of Bimastos and *Eophila* as *Allolobophora*, *Dendrobaena*, or *Eisenia*.

- the first critical use of the somatic characters of body pigmentation and setal arrangement in an earthworm classification scheme.

Pop was the first to consider the structure of muscle fibres in the longitudinal musculature, but he did not employ this character consistently throughout his taxonomic system. He did establish a new basic taxonomic system for lumbricid earthworms which correctly classified a large number of species.

Notwithstanding this valuable contribution, Pop did place into the genus *Allolobophora* species that belong to different genera and subgenera. The important result of Pop's system was that it introduced a new approach to solving taxonomic problems regarding earthworms, by forcing authors to re-evaluate old taxonomic characters and consider new ones.

It was almost 15 years before any major new effort was made to revise Pop's system. Omodeo (1956) made an attempt at a general solution to the taxonomic classification of earthworms, by employing new characters in the diagnosis of genera. He took an interesting approach to the study of generic biogeography, by comparing present distribution with paleogeographical maps. On the basis of these comparisons, he determined the distribution types of a single species. Omodeo's solution to the taxonomic classification of species was based on the available data at the time. Currently, we attribute species he classified as *Eophila* in different genera. Omodeo introduced into earthworm taxonomy several new characters, e.g. shape of calciferous glands, chromosome number, and structure of longitudinal muscle fibres, etc.

Omodeo described a larger number of new genera and subgenera:
- he divided *Dendrobaena* into subgenera *Dendrobaena* and *Dendrodrilus*,
- he divided *Octolasion* into subgenera *Octolasion* and *Octodrilus*,
- he divided *Allolobophora* into subgenera *Allolobophora* (3 groups) and *Cernosvitovia*, and genera *Bimastos*, *Eiseniona*, *Eophila* (3 groups), *Helodrilus*, and *Microeophila*,
- he expanded the diagnosis of *Eisenia*, *Eiseniella*, and *Lumbricus*,
- he erected *Eiseniona* as a new genus, many of its species are now in *Aporrectodea*.

Gates (1957) produced the first of what would become by 1980 a 26 part series of *Contributions to a Revision of the Earthworm Family Lumbricidae*. This was followed in 1958 by part two. These 26 papers by Gates are the only serious attempt, to date, by a North American to revise the Lumbricidae. In 1968, Gates produced three more papers in his lumbricid revision series (Gates 1968a–c).

Gates (1969a) redefined two lumbricid earthworm genera, *Bimastos* and *Eisenoides*, the latter erected for two North American species which had for years been attributed to *Allolobophora*, *Eisenia*, and *Helodrilus*. This was his sixth paper in his lumbricid series, which was followed a few months later by his seventh (Gates 1969b).

The next major revision to lumbricid taxonomy was that of Bouché (1972). He retained a part of Omodeo's taxonomic system (genera *Dendrobaena*, *Octolasion*, *Octodrilus*, *Eiseniella*, *Lumbricus*) and retained the subgenera

Dendrobaena and *Dendrodrilus*. In addition, he described the genus *Kritodrilus*. He included in his *Eisenia* (sensu Pop), two species which have subsequently been placed in different genera (*Allolobophoridella eiseni* and *Bimastos parvus*).

Bouché made dramatic changes when it came to *Allolobophora* and *Eophila*. Based on species found in France, he classified them as *Allolobophora*, *Helodrilus* and the new genus *Nicodrilus* (now junior synonym of *Aporrectodea*) with subgenera *Nicodrilus* and *Rhodonicus*. The species of *Allolobophora* were classified as *Allolobophora* sensu stricto and sensu lato, which would indicate that he did not find a satisfactory solution to the classification of the species in this genus. The genus *Eophila* was replaced by his three new genera: *Orodrilus*, *Prosellodrilus*, and *Scherotheca*.

The following is a summary of Bouché's (1972) contribution to the revision of the Lumbricidae:

- descriptions of 24 new species, 28 subspecies, and 18 varieties, notwithstanding that the *International Code of Zoological Nomenclature* [Art. 45(a)] states that the species-group includes only the specific and subspecific categories,
- six new genera erected: *Orodrilus*, *Prosellodrilus*, and *Scherotheca* (to deal with problems in *Dendrobaena* and *Allolobophora*, plus the elimination of *Eophila*), *Ethnodrilus* (primarily for ecological reasons), *Kritodrilus* (to classify Tétry's species *calarensis*), and *Nicodrilus* (now = *Aporrectodea*, to help with the *Allolobophora* problems).
- four new subgenera erected,
- *Dendrodrilus* retained as a subgenus of *Dendrobaena*; species of *Bimastos*, *Eophila*, and *Helodrilus* of others placed in various genera, and
- large quantities of ecological data attributable to species; extensive sampling employed, plus access to numerous collections of colleagues.

In 1972, Gates continued with two more papers (8th and 9th) in his contribution series on lumbricid revision. The first was a large and significant contribution (Gates 1972a), wherein he demonstrated, using morphological somatic characters, that the trapezoides complex, or what was frequently a collection of up to seven species, with three to four frequently included in what is termed *Allolobophora caliginosa*, were in fact distinct species, i.e. *trapezoides* (Dugès 1828), *tuberculata* (Eisen 1874), *turgida* (Eisen 1873), *longa* (Üde 1885), *limicola* (Michaelsen 1890), *nocturna* (Evans 1946), and *icterica* (Savigny 1826). Gates stated early in his introduction what was his own "esotery", which was formulated initially during his early years of working with the classical earthworm taxonomic system. After working with earthworm systematics for 45 years, he summed it up as follows:

- All individuals that are closely related enough to belong to one species should demonstrate that relationship by a *considerable common anatomy* not subject to individual variation (disregarding abnormality and monstrosity).
- All species so closely related as to belong to a single genus and presumably with a common ancestry, should demonstrate that relationship by a *lesser amount of invariable common anatomy*.

- All genera belonging to a single family also should prove that relationship by a still *smaller amount of commonly shared, invariant anatomy*. (*Emphasis added*, Gates 1972a pp. 1–2.)

Gates' observation was that if one carefully examined classical and neoclassical definitions of genera and families, they would reveal a lack of this invariation. Two of the characters upon which Gates placed considerable emphasis in the separation of these species of the trapezoides complex were: location of genital tumescences, typhlosole termination, and segment number. Many European authors prefer to use *caliginosa* for *turgida* and several of the other species as forms or varieties of *caliginosa*, e.g., *trapezoides* and *tuberculata* (Bouché, Mršić, Perel, Sims, and Zicsi). Bouché (1972) used different form names, which were soon put into synonymy, but were continued by Easton (1983) in his list. For a period of time, there was also confusion concerning *Lumbricus terrestris* and *Allolobophora terrestris* (Savigny) and with *Aporrectodea longa*, particularly in England. The taxonomic and nomenclatural problems which occurred in Europe, in this case, have not been encountered in North America.

Gates (1972a) placed these species in the genus *Aporrectodea*, within two years after this paper appeared. In subsequent years, Reynolds (1975a, b, 1976 a–d, 1994b and others) using these criteria, was able to show in biogeographical surveys throughout North America, that not only were these separate species, but that they had distinct distributional patterns. In a very recent study, Bøgh (1993), using electrophoretic techniques, upheld the existence of these species (in part, for those which were included in his study). In his second paper of the year (1972b), Gates dealt with his own evaluation of *Eisenoides*, which was made possible by many wide ranging collections of specimens by the Tall Timbers staff.

Perel (1973) compared the nephridial bladders from Bouché's genera with those of the Balkans in Omodeo's *Eophila*, and determined that the different (opposite) orientation indicated that the archaic species of France and those of the Balkans have resulted from independent development. In this review, Perel stated that Bouché's taxonomic system possessed no criteria allowing one to state the generic affinities of species.

Although the greatest problem with Bouché's system was his overwhelming reliance on intuition, he was the first to examine intergenerically the solutions to the problems of phylogeny. He introduced "series of species", by means of which the relationships of the species from the same genus could be shown in an illustrative manner.

In 1973, Gates continued with his 10th paper, in which he examined *Octolasion*. This paper was strongly critical of neoclassical revisionists, particularly Omodeo (1956), for failing to be precise with invariant characters, especially those of the more conservative somatic systems. He presented new information on the vascular and excretory systems and down-played the usefulness of classifying the calciferous sacs in this genus.

In the following year, Gates (1974a, b) continued his series with contributions on *Dendrobaena octaedra* and *Eisenia rosea*. The first paper is significant, because of the discussion of the importance of polymorphism and parthenogenetic populations for the classification of earthworms. Gates drew on many years of experience in studying the pheretimoid earthworms (Megascolecidae), where parthenogenetic morphs in many of the hundreds of species had created a nightmare for oligochaetologists over the decades.

It was a busy year for Gates when, during 1975, he published six new papers in his contribution series on the revision of the Lumbricidae. In his first paper of the year (Gates 1975a), he described the genus *Satchellius* with *Enterion mammale* Savigny 1826 as the type. In this paper, Gates also included new somatic information to be included in new diagnoses for 13 genera of the Lumbricidae. The characters he included were: calciferous sacs, position of calciferous gland opening into esophagus, calciferous lamellae, nephridial vesicles, nephropores and, in some cases, pigmentation and prostomium. The other five papers (Gates 1975b–f) had no direct bearing on North American collections or distributions, except for a couple of references to U.S. Quarantine Bureau interceptions.

Bouché (1975) described a new genus, *Spermophorodrilus*, based on a collection from Albania. Traditionally, species belonging in this genus had been attributed to *Eophila* and *Bimastos*. Revisions by Omodeo (1989) and Mršić (1991) erected a new type and placed the original type designated by Bouché as a synonym for another Balkan subspecies, which may have been published in a journal unavailable to Bouché.

It was early 1976 when Gates officially published his transfer of *Eisenia rosea* to the genus *Aporrectodea*, giving his reasons after tracing the species path through various genera from 1826 to 1976. This was his 19th paper in the contribution to a revision of the family Lumbricidae (Gates 1976).

In 1976, Reynolds and Cook produced the original volume of *Nomenclatura Oligochaetologica—A Catalogue of Names, Descriptions and Type Specimens of the Oligochaeta*. This was the first time that the names of all oligochaetes, along with the citation of their descriptions and the location(s) of their type specimens, were assembled in one place. Three additional supplements have been produced, which update the original book (Reynolds and Cook 1981, 1989, 1993).

Perel (1976, 1979) evaluated numerous taxonomic characters in her generic revision, but restricted herself to the species found in the Soviet Union. One of the most important aspects of her study was the significance of the structure and shape of the nephridial bladders, and the type of muscle fibre arrangement in longitudinal muscles. She also considered her studies on ecotypes in the revision. Perel amassed considerable data from many sources in her attempt to revise *Allolobophora*; unfortunately, she restricted herself to Soviet species, although she had sufficient material for a complete revision. She retained *Dendrobaena, Dendrodrilus, Eisenia, Eiseniella, Octolasion*, and *Lumbricus*. She completed the diagnosis of the genus *Kritodrilus*. She divided *Allolobophora* (sensu Pop) into *Nicodrilus* (= *Aporrectodea*), *Allolobophora* (subgenera *Allolobophora* and *Svetlovia* [= *Perelia*]), and *Helodrilus*. *Perelia* was erected

by Easton (1983) as a new name for *Svetlovia* which had been used by Olga
Čekanovskaya in 1975 as a genus of Tubificidae.

Gates' 20th paper in his continual revision of the Lumbricidae was based on
the genus *Eiseniella* in North America (Gates 1977). In the subsequent year
(Gates 1978a, b), he followed with revisions of the genera *Lumbricus* and
Eisenia, respectively. In his paper on the genus *Lumbricus*, he included a key
to 14 genera of the Lumbricidae, based solely on somatic characters. It was
during this discussion that he erected the new genus *Murchieonia*, with
Allolobophora minima Muldal, 1952 (= *Bimastos muldali* Omodeo 1956) as type
species. This genus was erected on the combination of several unique gut
characters.

In 1978, Zicsi described the genus *Fitzingeria*, including species which were
previously placed by other lumbricologists in *Lumbricus* (Örley) and *Dendrobaena* (Rosa, Michaelsen and Perel).

It was 1979, when Gates (1979a) presented his revision of the genus *Dendrodrilus* (sensu Omodeo), including a discussion of the *subrubicunda* and *tenuis*
morphs. Gates considered *Dendrodrilus* to be monospecific, although a recent
study would indicate that the *subrubicunda* morph may be a true species (Bøgh
1993).

In late 1980, Gates produced the last two papers in his revision of the
Lumbricidae (Gates 1980a, b). The first was his revision of the genus
Allolobophora, as he searched for the invariant somatic characters which he felt
would "enable a solution of the 'systematic chaos' in the European portion of
the Lumbricidae." Based on digestive and excretory invariant somatic characters
(nephridial vesicles, nephropores and calciferous glands), he supported Omodeo
(1956) and subsequent authors, who accepted *Al. chlorotica* as type species
which had been left undesignated in the neoclassical revisionists from Beddard
through Michaelsen and Stephenson. His last paper, Gates (1980b) dealt with
two octolasian species which have not been recorded in North America.

Zicsi (1981) elevated the subgenus *Cernosvitovia* Omodeo, 1956 of *Allolobophora* to full generic rank. This has been supported by eastern European
lumbricologists ever since. None of the species are found outside eastern
Europe.

The centenary (1982) of the publication of Darwin's book *The Formation of
Vegetable Mould through the Action of Worms* was the occasion for a symposium
which attracted some 150 participants, to discuss various aspects of earthworm
ecology. Two papers are of particular interest to this discussion, Sims (1983)
and Easton (1983). Chapter 40 was Sims' paper, entitled *The Scientific Names
for Earthworms*. In his seven page paper, he devoted a page to the
Allolobophora problem mentioned earlier and the *caliginosa* confusion. Sims
also devoted considerable space to the *Octolasion* problems (orthography,
nomenclature, and taxonomy). In an attempt to obtain universality and remove
confusion in earthworm taxonomy, Easton (1983) produced a checklist of valid
lumbricid species through December 1981, based heavily on Bouché (1972),
Perel (1979), and Zicsi (1982).

Fender (1985) followed up his earlier paper (Fender 1982) with a systematic discussion of all eleven lumbricid genera found in the eleven-state area of the western United States. In this survey, Fender included considerable amounts of new material collected over the years by McKey-Fender and himself, along with a review of the literature available at the time. He devoted considerable space to evaluating the European and North American positions on *Allolobophora* and *Aporrectodea*. He recognized that Levinsen's *eiseni* did not belong in any as yet described genus, and did not agree with Bouché's (1972) or Gates' (1978a) positions. In his discussion of *Dendrodaena*, he had the opportunity to include two species not previously recorded in the West or widely in North America. Based on somatic characters (e.g., nephropores), Fender believed that neither *Dendrobaena* nor any other currently described genus is the proper place for this species. Fender was responsible for the major collections of species of *Octolasion* in the West, and these observations are included in his discussion.

Zicsi (1985) continued his revision of *Allolobophora* with the description of the new genus *Proctodrilus*, containing species primarily from eastern Europe, but one or two species extending as far west as France.

Mršić (1987a, b) described three new genera, viz. *Creinella*, *Meroandriella*, and *Alpodinaridella* with two subgenera, *Dinaridella* and *Alpodinaridella*. Most of these new taxa were based on specimens from areas in and adjacent to his native Slovenia. By 1991, Mršić had reduced *Creinella* and *Meroandriella* to subgenera of *Aporrectodea*.

Together with Šapkarev in 1988, Mršić revised the genus *Allolobophora* sensu Pop, based on 47 species found in the Balkans. They divided the species into 10 genera and 9 subgenera, including two new genera, *Italobalkaniona* and *Karpatodinariona*, with its two new subgenera, *Panoniona* and *Serbiona*. Again, these are primarily Balkan species. By 1991, Mršić had elevated *Panoniona* and *Serbiona* to full generic status.

Omodeo (1988) revised his diagnosis of *Eophila*, placing some of the species in Bouché's *Scherotheca*, on the basis of only one morphological character (number of spermathecae), and by area maps in which these genera ranged from Iberia to central Asia. Omodeo apparently did not consider newer taxonomic characters proposed by various authors in more recent times.

In the following year, Omodeo (1989) described the genus *Healyella*, and created two new subfamilies (Lumbricinae and Spermophorodrillinae). The latter was based on its lack of tubercula pubertatis and spermathecae, but the presence of a ridged 15th segment with a male aperture, similar to the tubercula pubertatis. These are new characteristics of earthworm structure and are related to the shifting of the male aperture posterior to the 15th segment.

In 1990, Mršić redefined his earlier revision of *Allolobophora* and included a new subgenus *Allolobophoridella*, with *Lumbricus eiseni* Levinsen, 1884 as the type. In 1991, he also included *Allolobophora parva* Eisen, 1874 and elevated this subgenus to full generic rank. However, he did have some reservations about whether some of the other species of *Bimastos* in North America should remain in *Bimastos* or be transferred into his new genus. His concerns centre on the presence of tubercula pubertata on some European species, i.e., *B. beddardi*, vis-à-vis those of North America where these structures are absent, together with the shape of the nephridial bladder and its orientation (anteriorly vs. posteriorly).

The circulation of Mršić's major monograph (1991) was delayed, due to political unrest in the former Yugoslavia. In this monograph, he recognized 25 genera and 231 species. A summary of his monograph is as follows:

- described 21 new species,
- elevated subgenera *Allolobophoriella*, *Panoniona*, and *Serbiona* to generic rank,
- retained *Microeophila* and *Spermophorodrilus* as genera,
- ignored *Eophila* and transferred species to various genera derived from *Allolobophora* and did not know what to do with some and classified them as *species incertae sedis* (species of uncertain taxonomic position),
- ignored *Eiseniona* Omodeo, 1956 and transferred the species to *Aporrectodea* and *Eisenia*,
- reduced *Creinella* to subgeneric rank,
- made a detailed analysis of some important taxonomic characters and recognized their importance in helping to deal with phylogenetic and taxonomic problems,
- transferred *Bimastos parvus* and *Lumbricus eiseni* to *Allolobophoridella*, but admitted that new evidence based on orientation of nephridial bladders may result in this new genus being a synonym of *Bimastos*.

Based on the concept of and in the spirit of Easton (1983), I present (in part) the checklist of Reynolds and Wetzel (1994) for the current species of Lumbricidae found in continental North America, using the hierarchical system of Reynolds and Cook (1993):

ORDER HAPLOTAXIDA
SUBORDER LUMBRICINA
Superfamily Lumbricoidea
Family Lumbricidae
Subfamily Lumbricinae

Allolobophora Eisen, 1874
1. Type species: *Enterion chloroticum* Savigny, 1826   nUSA;CAN
    (now = *Allolobophora chlorotica*)
    Other species: in North America, none.
Allolobophoridella Mršić, 1990
2. Type species: *Lumbricus eiseni* Levinsen, 1884   AK,OR,TN,WA
    Other species: in North America, none.
Aporrectodea Örley, 1885
3. Type species: *Lumbricus trapezoides* Dugès, 1828   widespread
    Other species:
4. *Ap. icterica* (Savigny 1826)   NY,ON

5. *Ap. limicola* (Michaelsen 1890)	GA,MA,NJ,OR,PA,WA
6. *Ap. longa* (Üde 1885)	AL,CA,nUSA;eCan,BC
7. *Ap. rosea* (Savigny 1826)	widespread
8. *Ap. tuberculata* (Eisen 1874)	widespread
9. *Ap. turgida* (Eisen 1873)	widespread

Bimastos Moore, 1893

10. Type species: *Bimastos palustris* Moore, 1895	DE,NJ,NC,PA,TN

Other species:

11. *B. beddardi* (Michaelsen 1891)	11 states
12. *B. gieseleri* (Üde 1895)	9 states (s)
13. *B. heimburgeri* (Smith 1928)	14 states
14. *B. longicinctus* (Smith et Gittins 1915)	9 states (s)
15. *B. parvus* (Eisen 1874)	20 states (s)
16. *B. tumidus* (Eisen 1874)	19 states (s)
17. *B. welchi* Smith 1917	IN,KS,MO
18. *B. zeteki* (Smith et Gittins 1915)	15 states

Dendrobaena Eisen, 1874

19. Type species: *Dendrobaena boeckii* Eisen, 1874	nUSA;CAN

(= *Enterion octaedra* Savigny, 1826) (now = *Dendrobaena octaedra*)
Other species:

20. *Db. attemsi* Michaelsen, 1902	OR,WA
21. *Db. pygmaea* (Savigny 1826)	CA

Dendrodrilus Omodeo, 1956

22. Type species: *Enterion rubidum* Savigny, 1826	nUSA;CAN

(now = *Dendrodrilus rubidus*)
Other species: in North America, none.

Eisenia Malm, 1877

23. Type species: *Enterion fetidum* Savigny, 1826	widespread

(now = *Eisenia foetida*)
Other species:

24. *E. hortensis* (Michaelsen 1890)	8 states
25. *E. zebra* (Michaelsen 1902)	CA

Eiseniella Michaelsen, 1900

26. Type species: *Enterion tetraedrum* Savigny, 1826	widespread

(now = *Eiseniella tetraedra*)
Other species: in North America, none.

Eisenoides Gates, 1969

27. Type species: *Allolobophora lönnbergi* Michaelsen, 1894	14 states (c)

(now = *Eisenoides lönnbergi*)
Other species:

28. *Es. carolinensis* (Michaelsen 1910)	13 states (se)

Lumbricus L., 1758

29. Type species: *Lumbricus terrestris* L., 1758	widespread

Other species:

30. *L. castaneus* (Savigny 1826)   14 states (n); BC;eCAN
31. *L. festivus* (Savigny 1826)   VT;BC,NB,ON,PQ
32. *L. rubellus* Hoffmeister, 1843   widespread

Murchieona Gates, 1978

33. Type species: *Bimastos muldali* Omodeo, 1956   IN,MI,TN
       (= *Allolobophora minima* Muldal, 1952 preoccupied)
    Other species: in North America, none.

Octolasion Örley, 1885

34. Type species: *Lumbricus terrestris lacteus*   widespread USA;eCAN
       Örley, 1881
          (= *Enterion tyrtaeum* Savigny, 1826; *O. gracile* Örley, 1885)
          (now = *Octolasion tyrtaeum* (Savigny, 1826))
    Other species:
35. *O. cyaneum* (Savigny 1826)   13 states; 4 prov.

Satchellius Gates, 1975

36. Type species: *Enterion mammale* Savigny, 1826 (now = *Satchellius*
       mammalis)   NJ
    Other species: in North America, none.

III.   Survey of North American Earthworm Biogeography

The first attempt to present a regional view of the earthworms present in
continental North America was by Reynolds (1975a, 1976a). During the past 20
years, there have been many advances and additional available data for many
regions (states and provinces) in North America. The results of this additional
data are presented in a recent paper (Reynolds 1994b), which includes the latest
earthworm taxonomic and distribution maps, together with a discussion of the
general concepts of biogeography.

A table of regional earthworm surveys was presented by Reynolds et al.
(1974). Since that time, this author and others have made wide-ranging
collections in North America, resulting in numerous publications of the
distribution of earthworms in North America. The following table (Table 1) and
the list of species (*supra*) should give the reader a good overall impression of
the biogeography of North American lumbricid earthworms. Figures 1 to 6 at
the end of this chapter present distribution maps for many of the lumbricid
species found in Alaska, Canada, and the continental United States.

IV.   Current Research and Priorities for Future Research

One of the major problems for earthworm taxonomy in North America has been
the paucity of scientists, all living in isolation. Many individuals have dabbled
in earthworm research, but their contributions can frequently be counted on one

Table 1.   Regional Earthworm Surveys in North America.

Region	Number of Lumbricid Species	Number of Units (%) Surveyed[1]	References(s)
Alabama	10	75	Reynolds (1994g)
Alberta	6	30	Scheu & McLean (1993)
Arkansas	18	29	Causey (1952, 1953)
Connecticut	15	100	Reynolds (1973c)
Delaware	13	100	Reynolds (1973a)
Florida[2]	13	85	Reynolds (1994e)
Georgia[3]	19	40	Reynolds (prep)
Illinois	19	45	Harman (1960)
Indiana	23	100	Reynolds (1994a)
Louisiana[4]	8	100	Harman (1952), Gates (1965, 1967)
Maryland	20	100	Reynolds (1974)
Massachusetts	18	100	Reynolds (1977a)
Michigan	20	64	Snider (1991)
Mississippi	10	77	Reynolds (1994f)
Missouri	14	27	Olson (1936)
Montana	8	14	Reynolds (1972a)
New Brunswick	13	100	Reynolds (1976c)
New York	20	47	Oslon (1940), Eaton (1942)
North Carolina	21	80	Reynolds (1994c)
Nova Scotia	15	100	Reynolds (1976b)
Ohio	22	63	Olson (1928, 1932)
Ontario	19	96	Reynolds (1977b)
Oregon	18	70	MacNab & McKey-Fender (1947) Fender (1985)
Prince Edward	12	100	Reynolds (1975b)
Quebec (south shore)	15	100	Reynolds (1976d)
(north shore)	18	86	Reynolds & Reynolds (1992)
Rhode Island	11	100	Reynolds (1973b)
Tennessee	26	100	Reynolds et al. (1974)
Virginia	23	77	Reynolds (1994d)
Washington	15	40	Altman (1936), Fender (1985), MacNab & McKey-Fender (1947)

[1]   Units are counties, districts, parishes, etc.

[2,3]   Ten and eleven species of exotic pheretimoid species not included for Florida and Georgia, respectively.

[4]   Tandy (1969) studied only the genus *Pheretima* (8 species recorded for Louisiana).

hand. For two promising scientists, tragic death came early in life, before their full impact could be realized (William Murchie and Richard Tandy).

Gordon Gates, the Dean of Earthworm Taxonomy and Systematics, died in 1987. Although he began publishing in 1926, it was not until the latter part of his career that he devoted much time to the earthworms of North America and the Lumbricidae.

Prior to 1950, there were Frank Smith (1885–1937) and Henry Olson (1928–1940), who made contributions to taxonomy and distribution, respectively. In the last few decades, Dorothy McKey-Fender and William Fender have been concentrating on the taxonomy and distribution of the native and exotic earthworm fauna of the west coast of North America, and Sam James on the endemic species of the southeastern and plains areas.

Since 1972, I have collected widely throughout North America, alone and with the aid of my colleagues at Tall Timbers. The results of these collections have been published as distributional data primarily, with minimal contribution to systematics *sensu stricto*. Reynolds and Cook (1976, 1981, 1989, 1993) in the *Nomenclatura Oligochaetologica* series have brought together in one source the necessary reference data for anyone involved in the taxonomy and nomenclature of earthworms.

V.   Research Imperatives

With this limited background, I suggest the following priority for research and funding efforts:

A.   Training of Taxonomists

For over two decades (Reynolds 1972b, Reynolds et al. 1974), I have stated that the scarcity of competent earthworm systematists/taxonomists was detrimental to ecologists and others. The normal institutions which employ these types of specialists and encourage their development, e.g. museums and Departments of Agriculture, have not done so in North America since Smith worked for the Illinois Natural History Survey in the early 1900's. The difficulty encountered by the Organizing Committee of this conference in obtaining sufficient people to cover all the systematic topics attests to the fact that this problem still exists. There must be a concerted effort to support this type of research, before we are left without competent specialists in our field.

B.   Parthenogenesis in Taxonomy

The other major exotic group of earthworms in North America (Megascolecidae, pheretimoid groups) has long been plagued with taxonomic problems, which resulted from widespread parthenogenesis in its species (Gates 1972c). There is parthenogenesis within the Lumbricidae. One study recently has shown that localized populations of *Octolasion tyrtaeum* (Jaenike et al. 1980, 1982, Jaenike and Selander 1985) exhibited parthenogenesis. Previously, taxonomic problems with some morphs of what is now *Dendrodrilus rubidus* may be attributable to parthenogenesis. This area of research needs more attention.

C.   Earthworm Surveys

From Table 1 in this text it is obvious, in spite of what has already been accomplished, that there are major areas of North America in which earthworm surveys are lacking. In certain areas where native species exist, there is the potential for discovery of new species. Any new species in the Lumbricidae will probably come from the native genera *Bimastos* and *Eisenoides*, or a new genus.

D.   Life Histories

It is amazing that for the nearly 8000 oligochaetes (Reynolds and Cook 1993), modern, updated, life history studies have been done on only a few species, i.e. less than twenty. Some of the information gathered on common Lumbricidae was done at a time when species lumping occurred, i.e. factors attributed to *caliginosa* which included several species and thus accounted for the range of data, vis-à-vis specific values for other species.

E.   Modern Techniques

One area which was considered for years, but only recently had any evidence to support its potential, is electrophoresis. Bøgh (1993) illustrated that certain species were different, i.e. *Aporrectodea tuberculata* is a distinct species, and demonstrated how to identify species from fragments. This area of research should be followed, and will probably help with areas of confusion.

Acknowledgments

I would like to thank Wilma M. Reynolds of the Oligochaetology Laboratory, Lindsay, Ontario for reviewing numerous early drafts of this manuscript and her helpful comments and suggestions. Also, during the Workshop, many colleagues offered useful suggestions to help improve the manuscript. Again, I am indebted to Susan Mantle of the CARMA Centre at Sir Sandford Fleming College, Lindsay, Ontario for producing the maps which appear in this text. Two colleagues in particular, William M. Fender and Samuel W. James, generously shared some of their unpublished distribution records so that these maps might reflect the most up-to-date picture of North American lumbricid distribution.

References

Altman, L.C. 1936. *Oligochaeta of Washington*. Univ. Wash. Publ. Biol. 4:1–137.

Beddard, F.E. 1895. *A monograph of the order of Oligochaeta*. Oxford, 769 pp.

Bøgh, P.S. 1993. Identification of earthworms (Lumbricidae): choice of method and distinction criteria. *Megadrilogica* 4:163–174.

Bouché, M.B. 1972. *Lombriciens de France écologie et systématics*. Inst. Nat. Rech. Agron., Paris, 671 pp.

Bouché, M.B. 1975. La reproduction de *Spermophorodrilus albanianus* nov. gen., nov. sp. (Lumbricidae). *Zool. Jb. Syst.* 102:111.

Černosvitov, L. 1935. Monographie der Tschechoslovakischen Lumbriciden. *Arch. Přír. Vyzk. Čech.* 19:1–86.

Causey, D. 1952. The earthworms of Arkansas. *Proc. Ark. Acad. Sci.* 5:31–42.

Causey, D. 1953. Additional records of Arkansas earthworms. *Proc. Ark. Acad. Sci.* 6:47–48.

Easton, E.G. 1983. A guide to the valid names of Lumbricidae (Oligochaeta). p. 475–485. In: J.E. Satchell (ed.) *Earthworm ecology from Darwin to vermiculture*. Chapman and Hall, London.

Eaton, T.H. 1942. Earthworms of the northeastern United States: a key, with distribution records. *J. Wash. Acad. Sci.* 32:242–249.

Eisen, G. 1874a. Om Skandinaviens Lumbricider. *Öfv. K. Vet. Akad. Förh.* 30:43–56.

Eisen, G. 1874b. New Englands och Canadas Lumbricider. *Öfv. K. Vet. Akad. Förh.* 31: 41–49.

Fender, W.M 1982. *Dendrobaena attemsi* in an American greenhouse, with notes on its morphology and systematic position. *Megadrilogica* 4:8–11.

Fender, W.M. 1985. Earthworms of the western United States. Part I. Lumbricidae. *Megadrilogica* 4:93–129.

Gates, G.E. 1957. Contribution to a revision of the earthworm family Lumbricidae. I. *Allolobophora limicola*. *Breviora, Mus. Comp. Zool. Harvard*. No. 81:1–14.

Gates, G.E. 1958. Contribution to a revision of the earthworm family Lumbricidae. II. Indian species. *Breviora, Mus. Comp. Zool. Harvard*. No. 91:1–16.

Gates, G.E. 1965. Louisiana earthworms. I. A preliminary survey. *Proc. La. Acad. Sci.* 28:12–20.

Gates, G.E. 1967. On the earthworm fauna of the Great American desert and adjacent areas. *Great Basin Nat.* 27:142–176.

Gates, G.E. 1968a. What is *Lumbricus eiseni* Levinsen, 1884 (Lumbricidae, Oligochaeta)? *Breviora, Mus. Comp. Zool. Harvard*. No. 299:1–9.

Gates, G.E. 1968b. Contributions to a revision of the Lumbricidae. III. *Eisenia hortensis* (Michaelsen, 1890). *Breviora, Mus. Comp. Zool. Harvard*. No. 300:1–12.

Gates, G.E. 1968c. What is *Enterion ictericum* Savigny, 1826 (Lumbricidae, Oligochaeta)? *Bull. Soc. Linn. Normandie 10th ser.* 9:199–208.

Gates, G.E. 1969a. On two American genera of the family Lumbricidae. *J. Nat. Hist.* 9: 305–307.

Gates, G.E. 1969b. Contributions to a revision of the earthworm family Lumbricidae. V. *Eisenia zebra* Michaelsen, 1902. *Proc. Biol. Soc. Wash.* 82:453–460.

Gates, G.E. 1972a. Toward a revision of the earthworm family Lumbricidae. IV. The trapezoides species group. *Bull. Tall Timbers Res. Stn.* No. 12:1–146.

Gates, G.E. 1972b. On American earthworm genera. I. *Eiseniodes* (Lumbricidae). *Bull. Tall Timbers Res. Stn.* No. 13:1–17.

Gates, G.E. 1972c. Burmese earthworms. An introduction to the systematics and biology of Megadrile oligochaetes with special reference to southeast Asia. Trans. *Am. Phil. Soc.* 62:1–326.

Gates, G.E. 1973. The earthworm genus *Octolasion* in America. *Bull. Tall Timbers Res. Stn.* No. 14:29–50.

Gates, G.E. 1974a. Contributions to a revision of the Lumbricidae. X. *Dendrobaena octaedra* (Savigny, 1826) with special reference to the importance of its parthenogenetic polymorphism for the classification of earthworms. *Bull. Tall Timbers Res. Stn.* No. 15:15–57.

Gates, G.E. 1974b. Contributions to a revision of the family Lumbricidae. XI. *Eisenia rosea* (Savigny, 1826). *Bull. Tall Timbers Res. Stn.* No. 16:9–30.

Gates, G.E. 1975a. Contributions to a revision of the earthworm family Lumbricidae. XII. *Enterion mammale* Savigny, 1826 and its position in the family. *Megadrilogica* 2(1):1–5.

Gates, G.E. 1975b. Contributions to a revision of the earthworm family Lumbricidae. XIII. *Eisenia japonica* (Michaelsen, 1891). *Megadrilogica* 2(4):1–3.

Gates, G.E. 1975c. Contributions to a revision of the earthworm family Lumbricidae. XIV. What is *Enterion terrestris* Savigny, 1826 and what are its relationships? *Megadrilogica* 2(4):10–12.

Gates, G.E. 1975d. Contributions to a revision of the earthworm family Lumbricidae. XV. On some other species of *Eisenia*. *Megadrilogica* 2(5):1–7.

Gates, G.E. 1975e. Contributions to a revision of the earthworm family Lumbricidae. XVII. *Allolobophora minuscula* Rosa, 1906 and *Enterion pygmaeum* Savigny, 1826. *Megadrilogica* 2(6):7–8.

Gates, G.E. 1975f. Contributions to a revision of the earthworm family Lumbricidae. XVIII. *Octolasion calarense* Tétry, 1944. *Megadrilogica* 2(7):1–4.

Gates, G.E. 1976. Contributions to a revision of the earthworm family Lumbricidae. XIX. On the genus of the earthworm *Enterion roseum* Savigny, 1826. *Megadrilogica* 2(12):4.

Gates, G.E. 1977. Contribution to a revision of the earthworm family Lumbricidae. XX. The genus *Eiseniella* in North America. *Megadrilogica* 3(5):71–79.

Gates, G.E. 1978a. The earthworm genus *Lumbricus* in North America. *Megadrilogica* 3(6):81–116.

Gates, G.E. 1978b. Contributions to a revision of the earthworm family Lumbricidae. XXII. The genus *Eisenia* in North America. *Megadrilogica* 3(8):131–147.

Gates, G.E. 1979a. Contributions to a revision of the earthworm family Lumbricidae. XXIII. The genus *Dendrodrilus* Omodeo, 1956 in North America. *Megadrilogica* 3(9):151–162.

Gates, G.E. 1979b. Contributions to a revision of the earthworm family Lumbricidae. XXIV. What is *Dendrobaena byblica* Rosa, 1893? *Megadrilogica* 3(10):175–176.

Gates, G.E. 1980a. Contributions to a revision of the earthworm family Lumbricidae. XXV. The genus *Allolobophora* Eisen, 1874, in North America. *Megadrilogica* 3(11):177–184.

Gates, G.E. 1980b. Contributions to a revision of the earthworm family Lumbricidae. XXVI. On two octolasia. *Megadrilogica* 3(12):205–211.

Harman, W.J. 1952. A taxonomic survey of the earthworms of Lincoln Parish, Louisiana. *Proc. La. Acad. Sci.* 15:19–23.

Harman, W.J. 1960. *Studies on the taxonomy and musculature of the earthworms of central Illinois.* Ph.D. dissertation. Champaign: Univ. Illinois, 107 pp.

Hoffmeister, W. 1845. *Die bis jetzt bekannten Arten aus der Familie der Regenwürmer.* Als Grundlage zu einer Monographie dieser Familie Brauschweig. F. Vieweg und Sohn, 43 pp.

Jaenike, J. and R.K. Selander. 1985. On the co-existence of ecologically similar clones of parthenogenetic earthworms. *Oikos* 44:512–514.

Jaenike, J., S. Ausubel, and D.A. Grimaldi. 1982. On the evolution of clonal diversity in parthenogenetic earthworms. *Pedobiologia* 23:304–310.

Jaenike, J., E.D. Parker, Jr., and R.K. Selander. 1980. Clonal niche structure in the parthenogenetic earthworm *Octolasion tyrtaeum. Am. Nat.* 116:196–205.

Linnaeus, C. 1758. *Systema Naturae. Regnum Animale* (10th ed.). 824 pp.

Malm, A.W. 1877. Om daggmasker, Lumbricina. Öfv. Salsk. Hortik. *Vanners Göteborg Förh.* 1:34–47.

Macnab, J.A. and D. McKey-Fender. 1947. An introduction to Oregon earthworms with additions to the Washington list. *Northwest Sci.* 21:69–75.

Michaelsen, W. 1900. Oligochaeta. In: *Das Tierreich.* Leif 10. Verlag von R. Friedländer und Sohn, Berlin.

Moore, H.F. 1893. Preliminary account of a new genus of Oligochaeta. *Zool. Anz.* 16:333.

Moore, H.F. 1895. On the structure of *Bimastos palustris*, a new oligochaete. *J. Morph.* 10:473–496.

Mršić, N. 1987a. *Alpodinaridella* gen. nov. (Lumbricidae) and description of two new monotypic genera. *Biol. Věstn.* 35:61–66.

Mršić, N. 1987b. Description of a new genus and five species of earthworms (Oligochaeta: Lumbricidae). *Scoploia* 13:1–11.

Mršić, N. 1990. Description of a new subgenus, three new species and taxonomic problems of the genus *Allolobophora* sensu Mršić and Šapkarev 1988 (Lumbricidae, Oligochaeta). *Biol. Věstn.* 38:49–68.

Mršić, N. 1991. *Monograph on earthworms (Lumbricidae) of the Balkans.* Slovenska Akademija Znanosti Umetnosti, Ljubljana, Slovenia. 757 pp.

Mršić, N. and J. Šapkarev. 1988. Revision of the genus *Allolobophora* Eisen 1874 (emend. Pop 1941) (Lumbri cidae, Oligochaeta). *Acta Mus. Mac. Sci. Nat.* 1:1–38.

Olson, H.W. 1928. The earthworms of Ohio, with a study of their distribution in relation to hydrogen-ion concentration, moisture and organic content of the soil. Bull. *Ohio Biol. Surv.* 4, Bull. 17:47–90.

Olson, H.W. 1932. Two new species of earthworms for Ohio. *Ohio J. Sci.* 32:192–193.

Olson, H.W. 1936. Earthworms of Missouri. *Ohio J. Sci.* 36: 102–193.

Olson, H.W. 1940. Earthworms of New York state. *Am. Mus. Nov.*, No. 1090, 9 p.

Omodeo, P. 1956. Contributo all revisione dei Lumbricidae. *Arch. Zool. It.* 41:129–212.

Omodeo, P. 1988. The genus *Eophila* (Lumbricidae, Oligochaeta). *Boll. Zool.* 55:73–84.

Omodeo, P. 1989. Earthworms of Turkey. *Boll. Zool.* 56:167–199.

Örley, L. 1882. A palearktitus ovben elo terrikolaknak revizioja es elterjedese. *Ertek. Term Magyar Akad.* 15(18):1–34.

Perel, T.S. 1973. The shape of nephridal bladders as a taxomonic character in the systematics of Lumbricidae. *Zool. Anz.* 191:310–317.

Perel, T.S. 1976. A critical analysis of the Lumbricidae genera system. *Rev. Écol. Biol. Sol* 13:635–643.

Perel, T.S. 1979. *Rasprostranenie i zakonomernosti raspredelenia doždevyh červej fauny SSSR.* [Range and regularities in the distribution of earthworms of the USSR fauna] Nauka, Moskva, Russia. 268 pp.

Pop, V. 1941. Zur Phylogenie und Systematik der Lumbriciden. *Zool. Jb. Syst.* 74:487–522.

Reynolds, J.W. 1972a. A contribution to the earthworm fauna of Montana. *Proc. Mont. Acad. Sci.* 32:6–13.

Reynolds, J.W. 1972b. Earthworm (Annelida, Oligochaeta) ecology and systematics. p. 95–120. In: Dindal, D.L. (ed.) *Proc. 1ˢᵗ Soil Microcommunities Conf.* Springfield: U.S. Atomic Energy Commn., Natl. Tech. Inform. Serv., U.S. Dept. Comm.

Reynolds, J.W. 1973a. The earthworms of Delaware (Oligochaeta: Acanthodrilidae and Lumbricidae). *Megadrilogica* 1(5):1–4.

Reynolds, J.W. 1973b. The earthworms of Rhode Island (Oligochaeta: Lumbricidae). *Megadrilogica* 1(6):1–4.

Reynolds, J.W. 1973c. The earthworms of Connecticut (Oligochaeta: Lumbricidae, Megascolecidae and Sparganophilidae). *Megadrilogica* 1(7):1–6.

Reynolds, J.W. 1974. The earthworms of Maryland (Oligochaeta: Acanthodrilidae, Lumbricidae, Megascolecidae and Sparganophilidae). *Megadrilogica* 1(11):1–12.

Reynolds, J.W. 1975a. Die biogeografie van Noorde-Amerikaanse (Oligochaeta) noorde van Meksiko— I. *Indikator* 7:11–20.

Reynolds, J.W. 1975b. The earthworms of Prince Edward Island (Oligochaeta: Lumbricidae). *Megadrilogica* 2(7):4–10.

Reynolds, J.W. 1976a. Die biogeografie van Noorde-Amerikaanse (Oligochaeta) noorde van Meksiko— II. *Indikator* 8:6–20.

Reynolds, J.W. 1976b. The distribution and ecology of the earthworms of Nova Scotia. *Megadrilogica* 2(8):1–7.

Reynolds, J.W. 1976c. A preliminary checklist and distribution of the earthworms of New Brunswick. *N.B. Nat.* 7:16–17.

Reynolds, J.W. 1976d. Catalogue et clé d'identification des lombricidés du Québec. *Nat. can.* 103:21–27.

Reynolds, J.W. 1977a. The earthworms of Massachusetts (Oligochaeta: Lumbricidae, Megascolecidae and Sparganophilidae). *Megadrilogica* 3(2):49–54.

Reynolds, J.W. 1977b. *The earthworms (Lumbricidae and Sparganophilidae) of Ontario.* Life Sci. Misc. Publ., Roy. Ont. Mus., 141 pp.

Reynolds, J.W. 1994a. The distribution of the earthworms (Oligochaeta) of Indiana: a case for the Post Quaternary Introduction Theory for megadrile migration in North America. *Megadrilogica* 5(3):13–32.

Reynolds, J.W. 1994b. The distribution of earthworms (Annelida, Oligochaeta) in North America. p. 133–153. In: Mishra, P.C., N. Behera, B.K. Senapati, and B.C. Guru (eds.) *Advances in ecology and environmental sciences.* Ashish Publication, New Delhi.

Reynolds, J.W. 1994c. Earthworms of North Carolina (Oligochaeta: Acanthodrilidae, Komarekionidae, Lumbricidae, Megascolecidae, Ocnerodrilidae and Sparganophilidae). *Megadrilogica* 5(6):53–72.

Reynolds, J.W. 1994d. Earthworms of Virginia (Oligochaeta: Acanthodrilidae, Komarekionidae, Lumbricidae, Megascolecidae and Sparganophilidae). *Megadrilogica* 5(8):77–94.

Reynolds, J.W. 1994e. Earthworms of Florida (Oligochaeta: Acanthodrilidae, Eudrilidae, Glossoscolecidae, Lumbricidae, Megascolecidae, Ocnerodrilidae, Octochaetidae and Sparganophilidae). *Megadrilogica* 5(12):125–141.

Reynolds, J.W. 1994f. Earthworms of Mississippi (Oligochaeta: Acanthodrilidae, Lumbricidae, Megascolecidae, Ocnerdrilidae and Sparganophilidae). *Megadrilogica* 6(3):21–33.

Reynolds, J.W. 1994g. Earthworms of Alabama (Oligochaeta: Acanthodrilidae, Eudrilidae, Lumbricidae, Megascolecidae, Ocnerodrilidae and Sparganophilidae). *Megadrilogica* 6(4):35–47.

Reynolds, J.W. and D.G. Cook. 1976. *Nomenclatura Oligochaetologica*, a catalogue of names, descriptions and type specimens of the Oligochaeta. Fredericton: University of New Brunswick, 217 pp.

Reynolds, J.W. and D.G. Cook. 1981. *Nomenclatura Oligochaetologica Supplementum Primum*, a catalogue of names, descriptions and type specimens of the Oligochaeta. Fredericton: University of New Brunswick, 39 pp.

Reynolds, J.W. and D.G. Cook. 1989. *Nomenclatura Oligochaetologica Supplementum Secundum*, a catalogue of names, descriptions and type specimens of the Oligochaeta. New Brunswick Mus. Monogr. Ser. (Nat. Hist.) No. 8, 37 pp.

Reynolds, J.W. and D.G. Cook. 1993. *Nomenclatura Oligochaetologica Supplementum Tertium*, a catalogue of names, descriptions and type specimens of the Oligochaeta. New Brunswick Mus. Monogr. Ser. (Nat. Hist.) No. 9, 39 pp.

Reynolds, J.W. and K.W. Reynolds. 1992. Les vers de terre (Oligochaeta: Lumbricidae et Sparganophilidae) sur la rive nord du Saint-Laurent (Québec). *Megadrilogica* 4(9):145–161.

Reynolds, J.W. and M.J. Wetzel. 1994. North American Megadriles (continental North America north of Mexico). In: Coates, K.A., S.R. Gelder, J. Madill, J.W. Reynolds, and M.J. Wetzel. (eds.) *Common and scientific names of aquatic invertebrates from the United States and Canada*. Annelida. Am. Fish. Soc. Spec. Publ. No. 17 (in press).

Reynolds, J.W., E.E.C. Clebsch, and W.M. Reynolds. 1974. The earthworms of Tennessee (Oligochaeta). I. Lumbricidae. *Bull. Tall Timbers Res. Stn.* No. 17:1–133.

Rosa, D. 1893. Revisione dei Lumbricidi. *Mem. Acc. Torino.* ser. 2, 43:399–476.

Savigny, J.C. 1826. Analyse d'un Mémoire sur les Lombrics par Cuvier. *Mém. Acad. Sci. Inst. France* 5:176–184.

Scheu, S. and M.A. McLean. 1993. The earthworm (Lumbricidae) distribution in Alberta (Canada). *Megadrilogica* 4(11):175–180.

Sims, R.W. 1983. The scientific names of earthworms. p. 467–474. In: Satchell, J.E. (ed.) *Earthworm ecology from Darwin to vermiculture*. Chapman and Hall, London.

Smith, F. 1917. North American earthworms of the family Lumbricidae in the collections of the United States National Museum. *Proc. U.S. Natl. Mus.* 52:157–182.

Snider, R.M. 1991. Checklist and distribution of Michigan earthworms. *Mich. Academician* 24:105–114.

Stephenson, J. 1930. *The Oligochaeta*. Clarendon Press, Oxford. 978 pp.

Svetlov, P. 1924. Beobachtungen über die Oligochäten des Gouv. Perm. I. Zur Systematik, Fauna und Ökologie der Regenwürmer. *Bull. Inst. Rech. Biol. Perm.* 2:315–328.

Tandy, R.E. 1969. *The earthworm genus Pheretima Kinberg, 1866 in Louisiana*. Ph.D. dissertation. Baton Rouge: Louisiana State Univ., 155 pp.

Zicsi, A. 1978. Revision der Art *Dendrobaena platyura* (Fitzinger, 1899) (Oligochaeta: Lumbricidae). *Acta Zool. Hung.* 24:439–449.

Zicsi, A. 1981. Probleme der Lumbriciden-Systematik sowie die Revision zweier Gattungen. *Acta Zool. Hung.* 27:431–442.

Zicsi, A. 1982. Verzeichnis der bis 1971 beschriebenen und revidierten Taxa der Familie Lumbricidae (Oligochaeta). *Acta Zool. Hung.* 28:421–454.

Zicsi, A. 1985. Über die Gattungen *Helodrilus* Hoffmeister, 1845 und *Proctodrilus* gen. n. (Oligochaeta: Lumbricidae). *Acta Zool. Hung.* 31:275–289.

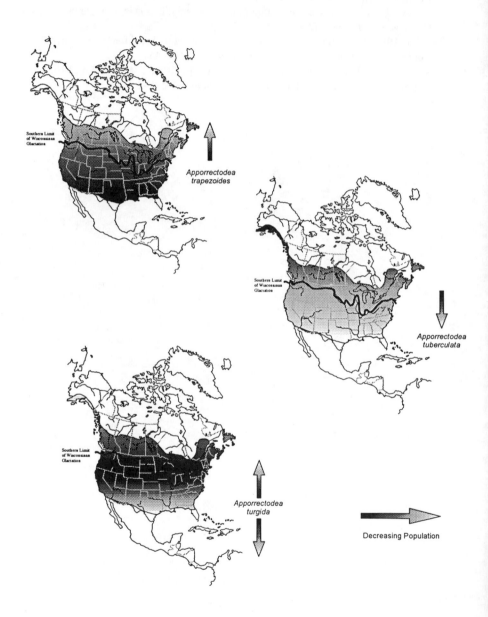

Figure 1. The approximate North American distribution of *Aporrectodea trapezoides*, *Ap. tuberculata*, and *Ap. turgida*.

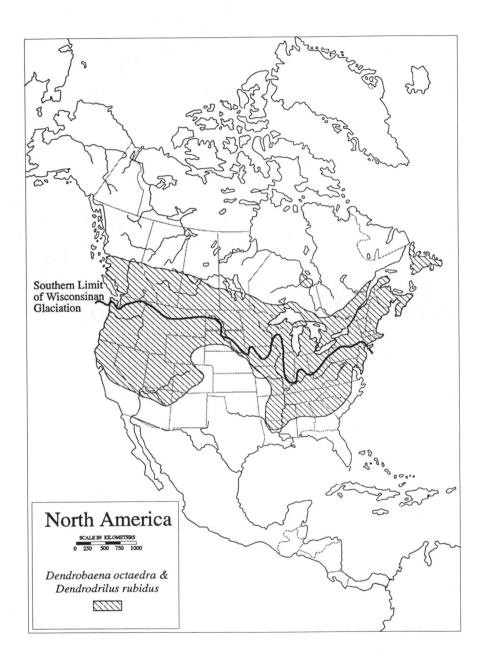

Figure 2. The approximate North American distribution of *Dendrobaena octaedra* and *Dendrodrilus rubidus*.

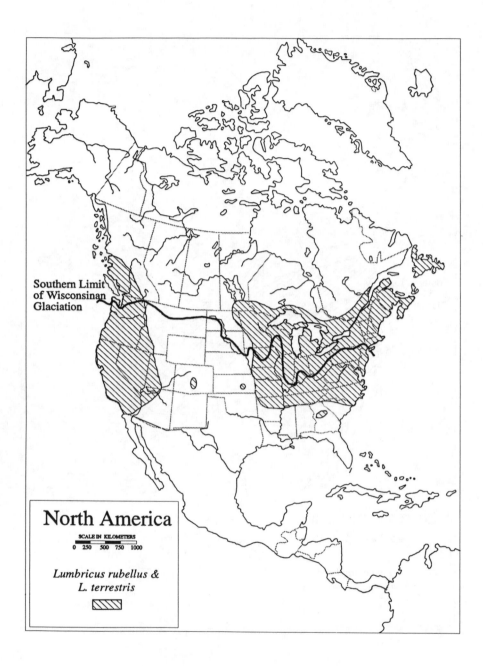

Figure 3. The approximate North American distribution of *Lumbricus rubellus* and *L. terrestris*.

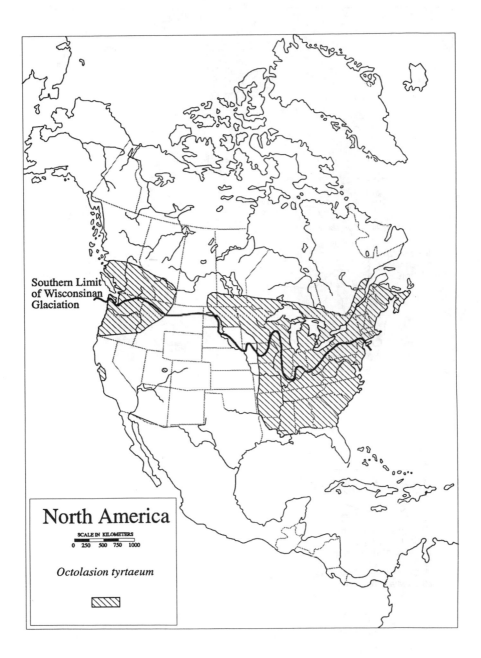

Southern Limit
of Wisconsinan
Glaciation

North America

SCALE IN KILOMETERS
0   250   500   750   1000

Octolasion tyrtaeum

Figure 4. The approximate North American distribution of *Octolasion tyrtaeum*.

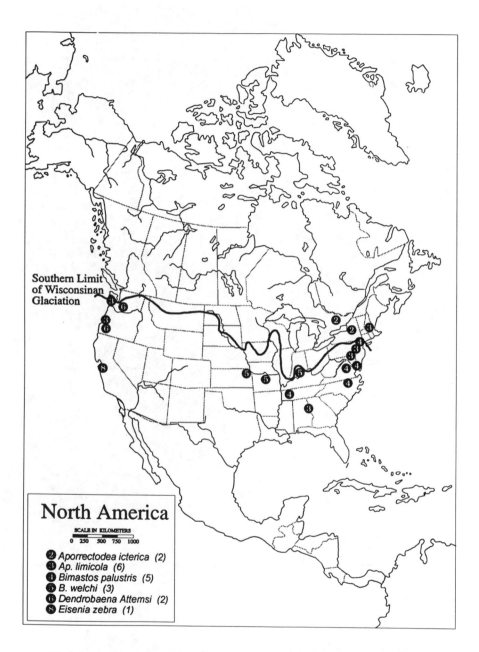

Figure 5. The limited North American distribution of some miscellaneous lumbricids (*Aporrectodea icterica*, *Ap. limicola*, *Bimastos palustris*, *B. welchi*, *Dendrobaena attemsi*, and *Eisenia zebra*). (The numbers in parentheses represent the number of collection sites.)

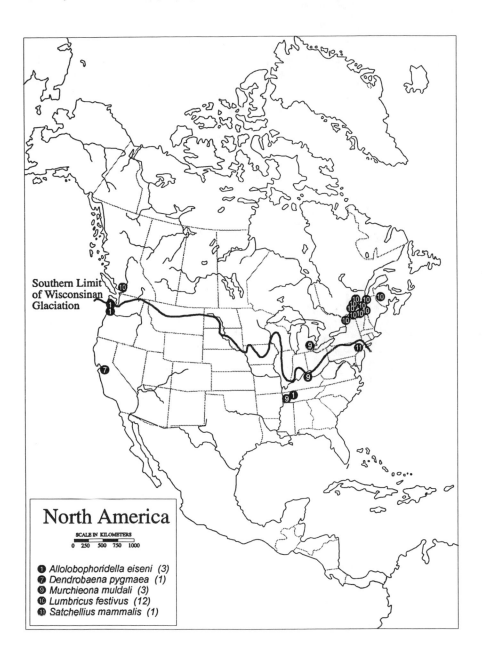

Figure 6. The limited North American distribution of some miscellaneous lumbricids (*Allolobophoridella eiseni*, *Dendrobaena pygmaea*, *Lumbricus festivus*, *Murchieona muldali*, and *Satchellius mammalis*). (The numbers in parentheses represent the number of collection sites.)

Systematics, Biogeography, and Ecology of Neartic Earthworms from Eastern, Central, Southern, and Southwestern United States

Samuel W. James

I. Introduction

In the region of the United States east of the continental divide and near to or south of the southern limit of Wisconsinan glaciation, plus the coastal mountains of California south of Riverside and Santa Ana, there are native representatives of five (possibly six) earthworm families (Gates 1982, James 1990a). The five are Lumbricidae (*Bimastos*, *Eisenoides*), Megascolecidae (*Diplocardia*), Sparganophilidae (*Sparganophilus*), Komarekionidae (*Komarekiona*), and Lutodrilidae (*Lutodrilus*). The sixth is the Ocnerodrilidae, unidentified members

1-56670-053-1/95/$0.00 + $.50

of which have turned up in southern California stream beds (Wood and James 1993) and in North Carolina (James, unpub.). In this paper I will attempt to summarize the current state of knowledge of these groups with respect to their systematics, biogeography, ecology, and economic importance. The task may not be very lengthy, since little is known of the lives of these animals. Systematics and biogeography will be considered together. I will not go into details on analytical aspects of biogeography, but instead focus on ranges of the groups. Range maps are provided at the end of this chapter (Figures 1 to 7), but one must bear in mind that data on ranges of most species are sparse. After concluding the systematics and biogeography section with recommendations for future activity in these fields, I will summarize research to date on biodiversity conservation, ecology, and economic applications.

II.   Survey of North American Native Earthworms

The organization (if not the content) of the above genera has been relatively stable for many years. One can imagine some reasons for this: the state of knowledge is complete and the relationships are so clear that there is no further reason to alter the system, or there has been so little work done on these worms that the current system's weaknesses have not been exposed. In some cases, the former is probably true, though one must add that new discoveries could change that. This is obviously true of the monotypic genera *Lutodrilus* and *Komarekiona*, for each of which a monotypic family has been defined. It may also be true of the monogeneric Sparganophilidae.

A.   Lutodrilidae

Lutodrilus is known only from the section of Louisiana between Baton Rouge and Bogalusa, where it can be quite abundant in riparian areas and mud flats (McMahan 1976, Harman pers. comm.). I am not aware of any attempts to find earthworms in similar habitats along the Gulf Coastal Plain or in other New World mud flat sites to the south. *Lutodrilus* probably occurs in adjacent portions of Mississippi, but it is not clear if anyone has looked. McMahan (1976) reports accounts of unusual earthworm-like oligochaetes in the Mississippi valley and along the Gulf coast, but nothing resembling the animals has been collected and preserved for modern observation. Like fantastic reports of giant reptiles lurking in swamps of equatorial Africa, verification of the existence of other Lutodrilidae awaits a mud-loving and intrepid explorer.

The Lutodrilidae are related to the Almidae, Biwadrilidae, and Sparganophilidae, in order of decreasing similarity (Jamieson 1988). Jamieson's analysis does not support affinity of Lutodrilidae with the lumbricoid families Lumbricidae, Glossoscolecidae, Hormogastridae, Komarekionidae, etc., as was thought to be the case by McMahan (1976). The Lutodrilidae and families similar to it are aquatic or limicolus (mud-inhabiting). The Almidae occur in Europe, Africa, and Asia, with one record in Central America, the Biwadrilidae are unique to Japan, and the Sparganophilidae are North and Central American. Because the ancestors of opisthoporous oligochaetes were probably aquatic or semi-aquatic, it should not be surprising to find a globally-distributed higher taxon of aquatic to limicolous families (cf. Jamieson 1988).

B.   Sparganophilidae

The Sparganophilidae are known from the entire geographic area covered in this chapter, except for the desert Southwest. However, it is probable that no one has ever looked for them there. The only genus, *Sparganophilus*, contains 13 species, a key to which is in Reynolds (1980). Two species are known from northern California and Oregon, and the rest from central, eastern, and southern states, plus the Canadian province of Ontario (detailed distribution data are in Reynolds 1975, 1977a, b, 1980). *Sparganophilus eiseni* has colonized aquatic and semi-aquatic habitats well to the north of the glacial boundaries (Smith and Green 1916, Reynolds 1977a). They may be found as lake benthos, in wet soils of stream banks or in saturated sediments along the edges of streams. New southern California records of an as yet undetermined *Sparganophilus* are in Wood and James (1993). The Sparganophilidae are most closely related to the Biwadrilidae, Lutodrilidae, and Almidae (Jamieson 1988), though classically they have been placed in or near the Glossoscolecidae (Michaelsen 1900, Stephenson 1930).

Species delineations within *Sparganophilus* are fairly clear, though the differences are often subtle, and many species descriptions are based on examination of very few specimens. The range of the most well-known species, *S. eiseni*, encompasses the ranges of its congeners except those of the Pacific region. In this case, one must be careful to determine if the range of morphological variation within *S. eiseni* is so narrow as to exclude the other species of the genus. Future research on *Sparganophilus* should focus on expanding collections of known species, searching unexplored areas, and otherwise clarifying the relationships among the various species.

C.   Komarekionidae

The range of *Komarekiona eatoni* is in deciduous forests of the eastern United States, from southern Pennsylvania south to North Carolina and Tennessee, and west into southern Indiana and Illinois (Gates 1974, 1982, Dotson and Kalisz 1989, James unpub.) However, Gates did not consider the Illinois and Indiana records to be in the natural range of the species. Though Gates (1974) created the Komarekionidae for this species, Sims (1980) has argued that it belongs in the Ailoscolecidae. Jamieson (1988) supports the family Komarekionidae, his analysis showing it as sharing a common ancestor with all other Lumbricoidea. Therefore, *Komarekiona* is of some systematic interest in determining the relationships within the superfamily and between the Lumbricoidea and other superfamilies.

The two remaining families with members clearly indigenous to temperate North America are the Lumbricidae and the Megascolecidae. Both are represented by many species and both are widely distributed in terrestrial habitats.

D.   Lumbricidae

Indigenous North American Lumbricidae fall into two genera, *Eisenoides* and *Bimastos* [see Schwert (1990) for a key]. The former is exclusively Nearctic and contains only two species so far. The latter is more diverse and has been assigned some species known only from Europe (Gates 1972a). *Eisenoides carolinensis* is found in mesic deciduous forested areas of the eastern United States, from the southern half of Pennsylvania to the Carolinas (Gates 1972b). It can be quite abundant locally. *Eisenoides lonnbergi* has approximately the same range, but extends farther north nearly to Lake Ontario. It inhabits wet areas, such as stream banks or stream beds, acid bogs, chronically damp soils and calcareous fens (pH range 4–7.8!). Generally it is found within forested areas, including coniferous forests of hemlock and white pine. It is probably the most northerly-distributed native earthworm of eastern North America, except for *S. eiseni*.

The genus *Bimastos* is generally found throughout the parts of eastern North America not glaciated in the Wisconsinan period and as far west as Kansas. It is also found in scattered locations in Michigan (Murchie 1956) and other areas slightly to the north of the glacial maxima. Most species are confined to forests where, with one exception (*B. longicinctus*), they live under the bark of decaying logs or in accumulations of organic matter such as leaf packs at the bases of slopes or near small drainages. *B. tumidus* has been observed to colonize rotting bales of straw in southeast Iowa and to be transported when the straw is moved from one location to another. Some species have been transported to other continents (Smith 1917, Gates 1972a).

Bimastos longicinctus also inhabits forested areas and rarely grasslands, but lives in the mineral soil rather than in highly organic media. *B. welchi* lives in tallgrass prairie of Missouri and Kansas, and has been collected from a nearly barren rocky slope on an otherwise forested hillside in west central Missouri. It is abundant in the Flint Hills of Kansas.

Currently, ten species of *Bimastos* are recognized. Other new species have recently turned up in North Carolina and the northern Appalachians (James, unpub., Schwert, pers. comm.). Ignoring for now the relationships of the North American *Bimastos* to European congeners, there is little dispute over the systematics of the group. Gates (1982) commented that *Bimastos* is composed of "parthenogenetic morphs, some nine or more of which have been regarded as species." In relation to this important evolutionary/systematic issue (can clonal organisms be regarded as species?), there should be an examination of the evidence for the alleged near-universality of parthenogenesis in *Bimastos*. I have collected numerous *B. welchi* bearing spermatophores. In much less extensive collecting than what took place in Kansas, I have found spermatophores on many other *Bimastos* species. I suggest that experimental and genetic approaches to determining the mode of reproduction in these and other reputedly parthenogenetic earthworms will be more convincing than noting the lack of spermathecae.

Other substantive concerns about *Bimastos* are raised in the unpublished notes of Gates. It appeared to him that there are forms intermediate between certain species, and that some of the key characters used in identification may be unreliable. His work was heading in the direction of locating reliable internal characters for defining species and identification of them.

E.   Megascolecidae

The sixth earthworm genus indigenous to the geographic area considered here is *Diplocardia*. I am following the classification proposed by Jamieson (1971a, b) and placing it in the subfamily Acanthodrilinae of the Megascolecidae. The Acanthodrilinae as defined by Jamieson are not necessarily equivalent to the Acanthodrilidae of other authors, so one should check before assuming that a genus said to belong to the Acanthodrilidae belongs to the Acanthodrilinae.

The range of *Diplocardia* extends from the southern half of Pennsylvania west through parts of Ohio, extreme southern Michigan, the southern halves of Indiana, Illinois, Iowa, and Nebraska, and south to Florida, the Gulf Coast, and west central Texas, plus records from Jeff Davis County in southwest Texas (Gates 1977). Then there is a gap in the known distribution, except for a few sporadic records from feedlots in New Mexico and Arizona (Gates 1967b) until California south of the Los Angeles basin. In that region some species live in native grassland remnants at middle and high elevations (Wood and James 1993, James 1994a). These new species are related to *D. keyesi*, known only from northern Baja California, Mexico (Eisen 1896). There are other *Diplocardia* from Mexico, but these will be considered in Chapter 4.

Diplocardia is closely related to the other acanthodriline genera *Notiodrilus* and *Diplotrema*, in that all have avesiculate holonephridia and the acanthodriline arrangement of male genitalia. *Diplocardia* differs in having a gizzard extending through two segments rather than one. The *Diplocardia* gizzard appears as one unit with little or no dividing line at the insertion of the septum 5/6, which septum is thin and membranous. This is in contrast to the obvious thinning of the gizzard wall and constriction at the septa that occur in *Zapotecia* and an undescribed species of a new Mexican genus related to *Zapotecia* (James 1994b). These distinct modes of gizzard multiplication or enlargement indicate that a common ancestor of *Diplocardia* and *Zapotecia* is probably monogiceriate,

Table 1. The currently recognized names of North American earthworms from the area east of the Rocky Mountains.

Komarekionidae
Komarekiona eatoni Gates 1974
Lumbricidae
Bimastos beddardi Mich. 1894
B. ducis Stephenson 1933
B. gieseleri Ude 1895
B. heimburgeri Smith 1928
B. longicinctus Smith and Gittins 1915
B. palustris Moore 1895
B. parvus Eisen 1874
B. tumidus Eisen 1874
B. welchi Smith 1917
B. zeteki Smith and Gittins 1915
Eisenoides carolinensis
E. lonnbergi
Lutodrilidae
Lutodrilus multivesiculatus McMahan 1976
Sparganophilidae
Sparganophilus eiseni Smith 1980
S. gatesi Reynolds 1980
S. helenae Reynolds 1980
S. komareki Reynolds 1980
S. kristinae Reynolds 1980
S. meansi Reynolds 1980
S. pearsi Reynolds 1975
S. smithi Eisen 1896
S. sonomae Eisen 1896
S. tennesseensis Reynolds 1977
S. wilmae Reynolds 1980
Megascolecidae
Diplocardia: with calciferous lamellae
D. alabamana Gates 1977
D. alba Gates 1943
D. bivesiculata Murchie 1961
D. eiseni Mich. 1894
D. farmvillensis Gates 1977
D. floridana Smith 1924
D. gatesi Murchie 1965

D. komareki Gates 1977
D. macdowelli Murchi 1967
D. michaelseni Eisen 1899
D. minima Gates 1977
D. mississippiensis Smith 1924
D. pettibonae Gates 1977
D. udei Gates 1955
D. vaili Gates 1977
D. varivesicula Murchie 1966
Diplocardia with spermathecae in segment vii:
D. biprostatica Gates 1977
D. communis Garmann 1888
D. caroliniana Eisen 1899
D. fusca Gates 1943
D. glabra Gates 1967
D. hulberti James 1988
D. longa Moore 1905
D. meansi Gates 1977
D. nova Gates 1977
D. ornata Gates 1943
D. sandersi Gates 1955
D. singularis Ude 1893
D. sylvicola Gates 1977
Other species:
D. bitheca Gates 1977
D. conoyeri Murchie 1961
D. fuscula Gates 1968
D. gracilis Gates 1943
D. invecta Gates 1955
D. kansensis James 1990
D. keyesi Eisen 1896
D. longiseta Murchie 1963
D. riparia Smith 1895
D. rugosa James 1988
D. smithii MacNab and McKey-Fender 1955
D. texensis Smith 1924
D. verrucosa Ude 1895

as opposed to the scheme proposed in Pickford (1937).

The other members of the Acanthodrilinae occur in Mexico and Central America, Chile, Argentina, South Africa, New Zealand, Australia, Madagascar, and various subantarctic islands (Pickford 1937, Michaelsen 1900, Lee 1959, Jamieson 1971a, b). This distribution is essentially Gondwanan, with an extension into North America.

Eisen (1896, 1900) proposed several subgenera of *Diplocardia*, none of which has been used since. The criterion was location of the male pores, which is quite variable among species known then and is more variable among species known now. The usual male pore location is xix, unlike most Acanthodrilinae in which it is in xviii. In Gates (1977) the species are arranged in three groups, those with calciferous glands, those with three pairs of spermathecae, and "other" species. Of these, only the first one is probably a natural grouping. However, in Gates (1977), *D. koebeli* is erroneously assumed to have calciferous glands. The species with calciferous glands are found from Pennsylvania south to Florida and west along the Gulf Coast states into Texas. The calciferous glands are of one basic type, differing in their degree of development and in their segmental location. There is little reason to suspect that calciferous glands arose independently in more than one lineage of *Diplocardia*.

My preliminary analyses suggest that a second natural group would be those species with spermathecae in segment vii, which includes the species with three pairs of spermathecae and two others with one and two pairs. This group is found throughout the range of the genus, except the Californian and Mexican portions. The bulk of the species are southern and central.

The remainder of the *Diplocardia* consists of species with two pairs of spermathecae and no calciferous glands, but is diverse with respect to male pore location and several other characters such as segment of intestinal origin and configuration of the spermathecae. Most of these species are from the central states and south into Texas. The species with posterior displacements of the male pores are mostly from the western portion of this zone and from California/Baja California.

While there are no absolute criteria for splitting or lumping of genera, *Diplocardia* now contains a greater diversity of character states than most authors seem to tolerate in other genera. For example, few other genera in the Megascolecidae include species with and without calciferous glands. A great deal of the generic definitions in the Glossoscolecidae depend on numbers and kinds of calciferous glands, and absence would be a cause for elevation to generic rank in that family. Male pores range from xviii to xxii within the "remainder" category of *Diplocardia*, a diversity that may be unparalleled in any other earthworm genus. If *Diplocardia* has survived intact this long, it is probably because the species "look like *Diplocardia*" and no one has wanted to break up a natural-seeming group. The time may be coming, because many new species in all sections of the genus have been collected (James, unpub. data).

There are some species or species groups for which the current system of classification is unclear. One group is *D. singularis* and *D. caroliniana* and their subspecies. The distinguishing characteristics are few, their variability considerable, and at least one subspecies, *D. singularis fluviatilis* may merit species rank. It inhabits mudflats of the Illinois River, whereas other *D. singularis* are found in dryland soils. It is possible that some subspecies, or some of the other variation, is associated with parthenogenesis, as appears to be true of *D. s. egglestoni* (Murchie 1958).

Another trouble spot is developing with *D. communis*, its subspecies *wolcotti*, and *D. nova*. One of the characters used to distinguish among them is the extent of doubling of the dorsal blood vessel. Unfortunately this character is highly variable in specimens recently collected in Iowa and Missouri. It may be necessary to rethink the species definitions. More useful characters may exist, but have not been fully explored or incorporated into keys.

Apart from these there are hints of geographic variation within worms that key to *D. verrucosa*, a very distinctive variant of *D. alba* given subspecific rank as *gravida* (Gates 1977), and numerous morphs and variants mentioned in Gates (1977). This amounts to saying that there is quite a bit of work yet to be done. Considering that most of the range of the genus has not been surveyed except in a very spotty fashion, this is not surprising.

From the practical point of view, we need to make progress in making it possible for everyone working on North American earthworms to identify their species. Doing taxonomic work may be fun for some, but many people have only one desire: to get a name for the specimen. Fender and McKey-Fender (1990), James (1990a), and Schwert (1990) made some headway, but there are some simple steps I believe would be very effective. The first would be periodic publication or distribution of updated keys. The keys should be accompanied by illustrations and a brief diagnosis of each species. Second, a computerized identification system/database should be kept and updated annually or as necessary. The software for this exists now and the whole package could be distributed over a computer network, such as INTERNET. Finally, there needs to be more support for the taxonomic work that could be done on material presently in collections and for surveying of the little-explored areas of North America.

III.   Biodiversity and Conservation Concerns

What do we mean by biodiversity? Among ecologists it is commonly understood as having something to do with species richness and evenness. Recently people began using it in a broader sense as "something that ought to be saved." Some of the important elements of a program to preserve biotic diversity are (1) awareness of the existence of the taxa to be preserved, (2) knowledge of the requirements for their preservation, (3) knowledge of present and future threats to the persistence of the taxa, and (4) social and political weight behind the programs to accomplish preservation. With reference to the earthworm genera discussed here, there are serious deficiencies in the first three, and it may be difficult to achieve the fourth at all. We simply do not know how many species of earthworms exist in North America and we do not know the range of any species with any confidence.

For some species it is possible to make educated guesses at the requirements for preservation. For example, most *Bimastos* species will be satisfied with a tract of forest with fallen trees in various stages of decomposition. From collection records we can glean the typical habitat conditions, but these do not necessarily represent minimum requirements. Considering the potential influences of exotic species of earthworms on the natives in their natural habitats, even habitat protection may fail (Dotson and Kalisz 1989, Kalisz and Dotson 1989). Nevertheless, preservation of ecosystems will reserve habitat for many cuteness-disadvantaged organisms, and probably is the best bet.

Threats to an earthworm species generally come from habitat destruction and invasion of exotics, singly or in combination (Smith 1928, Lee 1961, Stebbings 1962, Ljungstrom 1972). To date there are no experimental data on competition between native and introduced species from any system. This should be a top research priority. I have attempted the experiment twice without success. In both cases there was difficulty getting an area free of exotic species in which to install the experiment. Worm-free zones can be created in the field, but only at the cost of mass use of poisons or use of sites so severely disturbed that the generality of any results could be questionable. One alternative is to seek sites in naturally worm-free habitat in glaciated regions. These are not difficult to find, but are generally quite north of the ranges of the native species. That may introduce other factors irrelevant to the question about the outcome of a competitive interaction within the natural range of the native species.

IV. Ecology of North American Species

A.   Population Studies

Very little is known of the population dynamics or demography of North American earthworms. Murchie (1960) obtained information on the natural history of *B. zeteki* and Vail (1972) on natural history and reproduction in *D. mississippiensis*. James (1988b) contains data on population dynamics in relation to controlled burning. Populations of *Diplocardia smithii* and *D. verrucosa* had positive responses to fire in tallgrass prairie, while exotic Lumbricidae declined. The differences are probably related to different temperature tolerances, the native species being able to remain active at higher soil temperatures. Data gathered during the burning study and other projects were used to analyze seasonal changes in size-class distributions and response to dung pats in several species of *Diplocardia* and *Bimastos welchi* (James 1992a, b, *B. welchi* was erroneously identified as *Octolasion cyaneum*). From the former it was possible to conclude that different species of *Diplocardia* reproduced in different seasons, and that *B. welchi* has a very short early spring breeding period. Further information on seasonality of reproduction can laboriously be extracted from collection records, but there is no other research specifically on this subject.

James and Cunningham (1989) compared gut contents of several species of *Diplocardia*, *B. welchi* (again incorrectly identified as *O. cyaneum*) and *Aporrectodea turgida* and related the results to predictions of ecological niche based on gut morphology (cf. Lee 1959). There were good relationships between the data and predictions, but the relationships were most clear within family. Some *Diplocardia*, particularly *D. kansensis*, clearly fed on surface litter, while others fed in the soil. *Bimastos welchi* fed deepest in the soil of all native species, in stark contrast to the feeding behavior of its forest congeners.

There are some studies relating earthworm species populations to habitat, microsite, and/or soil influences (Murchie 1956, James 1988, Dotson and Kalisz 1989, Boettcher and Kalisz 1991, Wood and James 1993). Most of these describe naturally-occurring situations and consequently do not definitively test hypotheses about causation. They do not answer questions about why the worms do not live in other places or about factors controlling population sizes. This should be taken as suggestion, not as criticism. It really makes little sense to charge into sophisticated experimental design without a basis for the elements of the design. These studies provide good starting points. Just as plant ecology made a transition from largely descriptive work to extensive experimentation, we should look towards the same.

B.   Community Ecology and Earthworms

There has been no published research on the relationships among North American native earthworms and other biota, such as soil fauna. An example of a potential research program would be to investigate the effects of earthworm activity on soil arthropods. One could hypothesize that pore size distribution changes could affect the population sizes or species composition of the arthropods. In another example, earthworm alterations of litter decomposition rates may dramatically alter the soil/litter faunal community. It is possible that there may be further effects on other groups of organisms, such as fungi, bacteria, and plants. However, there has been no research in these areas.

C.   Agricultural and Other Economic Applications of
        North American Earthworms

The majority of research (what little there is so far) on earthworm ecology in pasture and arable lands in North America has dealt with European Lumbricid-ae. An exception is Teotia et al. (1950), in which a worm identified as *Diplocardia riparia* was present in cropland. The authors noted some differences in the stability of casts produced by native and introduced species. Gates (1967a, 1968) obtained specimens of *D. glabra* and *D. fusca* from Louisiana agricultural fields. Parmelee et al. (1990) and Hendrix et al. (1992) reported the presence of *D. caroliniana* in no-tillage agricultural fields and *Diplocardia* spp. in pastures on the Georgia piedmont. *Diplocardia* may be more important in croplands of southern areas because exotic Lumbricidae may experience high temperature limitation. On the other hand, there may be *Amynthas* species (Megascolecidae; from Asia) capable of colonizing these same areas.

Bimastos tumidus can become established in compost piles. I have observed it to multiply rapidly in old straw and spoiled hay as well, and to spread from one pile to another. This species has the potential to be useful in organic waste management or composting. This has not been rigorously researched, but it seems that it does well at medium C/N ratios (cool temperatures of decomposition), compared to *Eisenia foetida*. In the future it may be possible to propagate *B. tumidus* for acceleration of decomposition without raising the problems associated with introducing an exotic species. Another advantage is its apparent ability to survive ordinary soil conditions in the cold continental climate (USDA hardiness zone 5) of Iowa. Thus once established in an area, it would persist without periodic expensive reintroductions.

This is one example for which some preliminary observations are available. With dozens more native species from which to choose, it is possible that more will turn out to have economic uses.

Several species of *Diplocardia* are currently collected commercially for bait from natural populations in Kansas, Missouri, and Florida. Still others are used on a smaller scale, principally the larger species (e.g., *D. mississippiensis* and *D. floridana*) of southern states. Worm gathering is regulated in the Apalachicola National Forest of Florida, where the industry is of some importance. In Kansas and Missouri there is no regulation to date, so it may not be easy to determine the economic importance of *D. riparia*, the target species (sold as river worms). Fishermen in Kansas favor river worms for summer fishing because they do not require refrigeration (vs. *Lumbricus terrestris*) and remain lively longer in warm summer waters.

V.   Research Imperatives

It is obvious from the foregoing discussion that deficiencies exist in our knowledge of North American native earthworms. This section identifies specific research directions aimed at overcoming these deficiencies.

A.   Biological Survey and Inventory

Much of the continental United States has not been systematically surveyed for native earthworms. Basic information on the presence and abundance, and habitat relationships of species in all five families are needed across these poorly studied regions. In particular, the Gulf Coast drainage systems may yield new species of *Diplocardia* and possibly of *Lutodrilus*. Work is also needed on the Sparganophilidae, including expansion of collections of known species and surveys of unexplored areas for new species.

B.   Systematics and Evolutionary Issues

A number of unresolved questions need further research:
- Is *Komarekiona eatoni* best placed within Ailoscolecidae? What is the relationship of Komarekionidae to other families within the Lumbricoidea?
- Is parthenogenesis nearly universal in *Bimastos*? The occurrence of spermatophores in many species suggests otherwise. Experimental and genetic studies are needed to determine mode of reproduction. There also is need to continue the unfinished work of Gates in locating reliable internal characteristics for defining and identifying *Bimastos* species.
- Is *Diplocardia* a "good" genus? Three distinct natural groupings occur, based on morphology and partly on biogeography. The diversity of character states is greater than in most genera within the Megascolecidae and related families, and is increasing as new species are found. Furthermore, classifications are unclear for several species and subspecies (e.g., *D. singularis*, *D. caroliniana*, *D. communis*, and *D. verrucosa*). A thorough analysis and reevaluation of the genus is needed.

To better serve the needs of the research community, taxonomic keys to native and exotic species need to be updated periodically, with improved illustrations and brief diagnoses. Computerized identification systems should be distributed on computer networks for wider accessibility and easier updating. Finally, more support is greatly needed for taxonomic work, both on existing collections and on new collections from unsurveyed areas of North America.

C.   Biodiversity and Conservation

There are serious deficiencies in our knowledge of abundance and distribution of native earthworm species, their habitat requirements, and present and future threats to their existence throughout North America. Further, it is not clear that there is sufficient social or political interest to foster conservation of earthworm biodiversity. Because the greatest threats to earthworm species are habitat alteration and invasion of exotics, current efforts toward management and preservation of whole ecosystems should be promoted, and, as a top priority, experimental research is needed to investigate competitive interactions between native and exotic earthworm species. The importance of the latter is almost completely unknown.

D. Ecology of North American Species

- Little is known about the life history (e.g., population dynamics and demography) of native earthworms. Studies are needed especially on seasonality of reproduction and feeding ecology (i.e., niche characteristics). Experimental work on relationships between populations and habitat, microsite, and soil influences would reveal controls on distribution and abundance.
- Interactions between native earthworms and other biota are poorly known. Studies of effects of native species on plants, microbes, and other soil fauna would be useful contributions in this area, and would aid our understanding of the role of native earthworms in ecosystem processes.

E. Applied Ecology and Economic Considerations

- Little attention had been paid to native earthworm species in agricultural soils in North American. *Diplocardia* species may be important in the southern U.S. because of their tolerance of higher temperatures compared to introduced lumbricids which dominate in agricultural areas of the midwest, and which have been shown to significantly influence soil processes there.
- Further research also may show an important role for species of *Bimastos* (e.g., *B. tumidus*) in vermicomposting, which is increasing in use in the U.S.
- Finally, "cottage" fish-bait industries based on native earthworms exist in several states in the U.S. Mostly, these rely on harvesting worms in forests but research on techniques for cultivating these species could increase their production and reduce the demand for exploitation in natural ecosystems.

References

Boettcher, C.E. and P.J. Kalisz. 1991. Single-tree influence on earthworms in forest soils in eastern Kentucky. *Soil Sci. Soc. Am. J.* 55:862–865.

Dotson, D.B. and P.J. Kalisz. 1989. Characteristics and ecological relationships of earthworm assemblages in undisturbed forest soils in the southern Appalachians of Kentucky, USA. *Pedobiologia* 33:211–220.

Eisen, G. 1896. Pacific Coast Oligochaeta II. Mem. *Calif. Acad. Sci.* 2:123–199.

Eisen, G. 1900. Researches in American Oligochaeta, with especial reference to those of the Pacific coast and adjacent islands. *Proc. Cal. Acad. Sci. 3rd Ser. Zoology* 2:85–276

Fender, W.F. and D. McKey-Fender. 1990. Oligochaeta: Megascolecidae and other earthworms from western North America. p. 357–378. In: Dindal, D.L. (ed.) *Soil biology guide*. John Wiley and Sons, New York.

Gates, G.E. 1967a. Louisiana earthworms. III. *Diplocardia glabra*, sp. n. (Acanthodrilidae, Oligochaeta, Annelida). *Proc. Louisiana Acad. Sci.* 30: 26–31.

Gates, G.E. 1967b. On the earthworm fauna of the Great American Desert and adjacent areas. *Great Basin Nat.* 27:142–176.

Gates, G.E. 1968. Louisiana earthworms. IV. *Diplocardia fusca Gates 1943.* (Acanthodrilidae, Oligochaeta, Annelida). *Proc. Louisiana Acad. Sci.* 31:18–22.

Gates, G.E. 1972a. Burmese earthworms. An introduction to the systematics and biology of megadrile oligochaetes with special reference to southeast Asia. *Trans. Am. Phil. Soc.* 62:1–326.

Gates, G.E. 1972b. Contributions to North American earthworms (Annelida). No. 4 On American earthworm Genera. I. Eisenoides (Lumbricidae). *Bull. Tall Timbers Res. Stn.* No. 13:1–17.

Gates, G.E. 1974. On a new species of earthworm in a southern portion of the United States. Contributions on North American Earthworms (Annelida) no. 9. *Bull. Tall Timbers Res. Stn.* No. 15:1–13.

Gates, G.E. 1977. More on the earthworm genus *Diplocardia. Megadrilogica* 3:1–47.

Gates, G.E. 1982. Farewell to North American megadriles. *Megadrilogica* 4:12–77.

Hendrix, P.F., B.R. Mueller, R.R. Bruce, G.W. Langdale, and R.W. Parmelee. 1992. Abundance and distribution of earthworms in relation to landscape factors on the Georgia Piedmont, U.S.A. *Soil Biol. Biochem.* 24:1357–1361.

James, S.W. 1988a. *Diplocardia hulberti* and *D. rugosa*, new earthworms (Annelida: Oligochaeta: Megascolecidae) from Kansas. *Proc. Biol. Soc. Washington* 101:300–307.

James, S.W. 1988b. The postfire environment and earthworm populations in tallgrass prairie. *Ecology* 69:476–483.

James, S.W. 1990a. Oligochaeta: Megascolecidae and other earthworms from southern and midwestern North America. p. 379–386. In: Dindal, D.L. (ed.) *Soil biology guide,* John Wiley and Sons, New York.

James, S.W. 1990b. *Diplocardia kansensis*, a new earthworm from Kansas, with redefinitions of *D. riparia* Smith and *D. fuscula* Gates (Annelida: Oligochaeta: Megascolecidae). *Proc. Biol. Soc. Washington* 103:179–186.

James, S.W. 1992a. Localized dynamics of earthworm populations in relation to bison dung in North American tallgrass prairie. *Soil Biol. Biochem.* 24:1471–1476.

James, S.W. 1992b. Seasonal and experimental variation in population structure of earthworms in tallgrass prairie. *Soil Biol. Biochem.* 24:1445–1449.

James, S.W. 1994a. New species of *Diplocardia* and *Arigilophilus* (Oligochaeta: Megascolecidae) from southern California. *Proc. Biol. Soc. Washington.* (in press).

James, S.W. 1994b. New acanthodriline earthworms from Mexico (Oligochaeta: Megascolecidae). *Acta Zool. Mex.* (n.s.) 60:1–21.

James, S.W. and M.R. Cunningham. 1989. Feeding ecology of some earthworms in Kansas tallgrass prairie. *Am. Midl. Nat.* 121:78–83.

Jamieson, B.G.M. 1971a. A review of the Megascolecoid earthworm genera (Oligochaeta) of Australia. Part I. Reclassification and checklist of the megascolecoid genera of the world. *Proc. R. Soc. Qld.* 82:75–86.

Jamieson, B.G.M. 1971b. A review of the Megascolecoid earthworm genera (Oligochaeta) of Australia. Part II. The subfamilies Ocnerodrilinae and Acanthodrilinae. *Proc. R. Soc. Qld.* 82:95–108.

Jamieson, B.G.M. 1988. On the phylogeny and higher classification of the Oligochaeta. *Cladistics* 4:367–401.

Kalisz, P.J. and D.B. Dotson. 1989. Land-use history and the occurrence of exotic earthworms in the mountains of eastern Kentucky. *Am. Midl. Nat.* 122:288–297.

Lee, K.E. 1959. The earthworm fauna of New Zealand. *N.Z. Dept. Sci. Ind. Res. Bull.* *130*, Wellington, New Zealand.

Lee, K.E. 1961. Interactions between native and introduced earthworms. *Proc. N.Z. Ecol. Soc.* 8:60–62.

Ljungstrom, P.O. 1972. Taxonomical and ecological notes on the earthworm genus *Udeina* and a requiem for the South African acanthodrilines. *Pedobiologia* 12:100–110.

McMahan, M.L. 1976. Preliminary notes on a new megadrile species, genus, and family from the southeastern United States. *Megadrilogica* 2:6–8.

Michaelsen, W. 1900. Oligochaeta. In: *Das Tierreich*. Lief 10. Verlag von R. Friedländer und Sohn, Berlin.

Moore, J.P. 1895. On the structure of *Bimastos palustris*, a new oligochaete. *J. Morph.* 10:473–496.

Murchie, W.R. 1956. Survey of the Michigan earthworm fauna. *Pap. Mich. Acad. Sci., Arts, Let.* 41:53–72.

Murchie, W.R. 1958. A new megascolecid earthworm from Michigan with notes on its biology. *Ohio J. Sci.* 58:270–272.

Murchie, W.R. 1960. Biology of the Oligochaete *Bimastos zeteki* Smith and Gittins (Lumbricidae) in northern Michigan. *Am. Midl. Nat.* 64:194–215.

Parmelee, R.W., M.H. Beare, W. Cheng, P.F. Hendrix, S.J. Rider, D.A. Crossley Jr., and D.C. Coleman. 1990. Earthworms and enchytraeids in conventional and no-tillage agroecosystems: A biocide approach to assess their role in organic matter breakdown. *Biol. Fertil. Soils.* 10:1–10.

Pickford, G.E. 1937. *A monograph of the acanthodriline earthworms of South Africa*. Heffer, Cambridge. 612 pp.

Reynolds, J.W. 1975. *Sparganophilus pearsi* n. sp. (Oligochaeta: sparganophilidae) a nearctic earthworm from western North Carolina. *Megadrilogica* 2:9–11.

Reynolds, J.W. 1977a. *The earthworms (Lumbricidae and Sparganophilidae) of Ontario*. Life Sci. Misc. Publ., Royal Ontario Museum, Toronto.

Reynolds, J.W. 1977b. The earthworms of Tennessee (Oligochaeta). II. Sparganophilidae, with a description of a new species. *Megadrilogica* 3:61–64.

Reynolds, J.W. 1980. The earthworm family Sparganophilidae (Annelida: Oligochaeta) in North America. *Megadrilogica* 3:189–204.

Schwert, D.P. 1990. Oligochaeta: Lumbricidae. p. 341–356. In: Dindal, D.L. (ed.) *Soil biology guide*. John Wiley and Sons, New York.

Sims, R.W. 1980. A classification and the distribution of earthworms, suborder Lumbricina (Haplotaxida: Oligochaeta). *Bull. Brit. Mus. Nat. Hist. (Zool.)* 39:103–124.

Smith, F. 1895. A preliminary account of two new oligochaetes from Illinois. *Ill. State Lab. Nat. Hist. Bull.* 4:142–147.

Smith, F. 1917. North American earthworms of the family Lumbricidae in the collections of the United States National Museum. *Proc. U.S. Natl. Mus.* 52:157–182.

Smith, F. 1928. An account of changes in the earthworm fauna of Illinois and a description of one new species. *Ill. Nat. Hist. Surv. Bull.* 17:347–362.

Smith, F. and E.M. Gittins. 1915. Two new species of Lumbricidae from Illinois. *Ill. State Lab. Nat. Hist. Bull.* 10:545–550.

Smith, F. and B.R. Green. 1916. The Porifera, Oligochaeta, and certain other groups of invertebrates in the vicinity of Douglas Lake, Michigan. *Mich. Acad. Sci. Ann. Rept.* 17:81–84.

Stebbings, J.H. 1962, Endemic-exotic earthworm competition in the American Midwest. *Nature* 196:905–906.

Stephenson, J. 1930. *The Oligochaeta*. Clarendon Press, Oxford.

Teotia, S.P., F.L. Duley, and T.M.McCalla. 1950. Effect of stubble mulching on number and activity of earthworms. *Neb. Agr. Exp. Stn. Res. Bull.* 165:1–20.

Vail, V.A. 1972. Contributions to North America earthworms (Annelida). No. 1 Natural history and reproduction in *Diplocardia mississippiensis* (Oligochaeta). *Bull. Tall Timbers Res. Stn.* 11:1–39.

Wood, H.B. and S.W. James. 1993. Native and introduced earthworms from selected chaparral, woodland, and riparian zones in southern California. *USDA Forest Service Gen. Tech. Rep.* PSW-GTR-142.

Figure 1. The approximate North American distribution of *Komarekiona eatoni* and *Lutodrilus multivesiculatus*.

Figure 2. The approximate North American distribution of *Sparganophilus* spp.

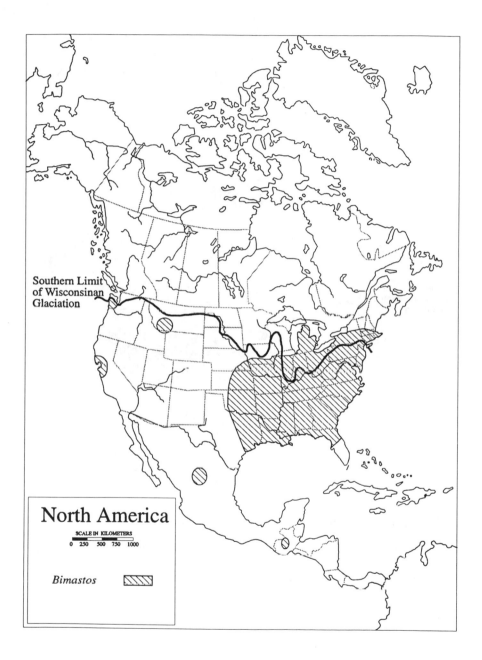

Figure 3. The approximate North American distribution of *Bimastos* spp.

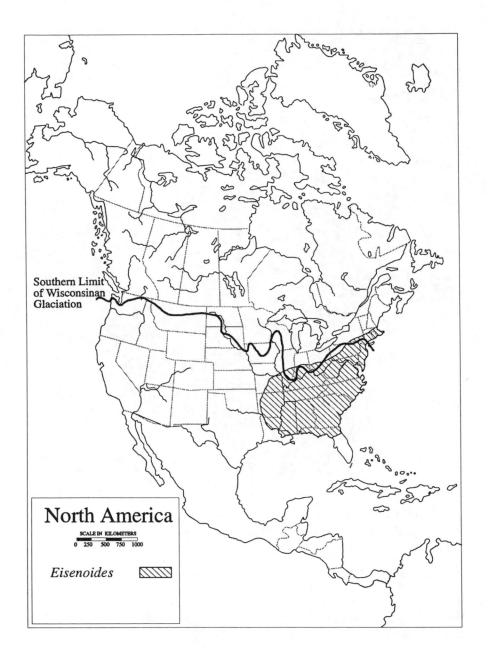

Figure 4. The approximate North American distribution of *Eisenoides* spp.

Figure 5. The approximate North American distribution of *Diplocardia* spp. with calciferous lamellae.

Figure 6. The approximate North American distribution of *Diplocardia* spp. with spermathacae in segment vii.

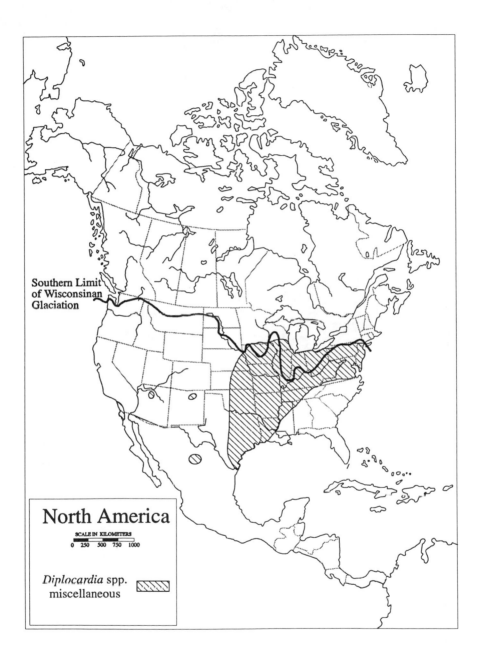

Figure 7. The approximate North American distribution of miscellaneous *Diplocardia* spp.

Native Earthworms
of the Pacific Northwest:
An Ecological Overview
William M. Fender

I.   Introduction

The earthworm fauna of the western United States is rich, varied, and interesting, but highly underreported. The most recent overview of the group (Fender and McKey-Fender 1990) puts the fauna at 9 genera and 28 species, but according to our estimates there are over 80 additional species in our collections waiting to be described. The geographic regions most needing work are northern California and central Idaho, but there are undescribed species from all over the region.

Until recently, all western North American Megascolecidae were placed in two genera, *Plutellus* and *Megascolides*, which by the concepts of the time, were worldwide in distribution. *Argilophilus* Eisen 1893 was described to contain a few Californian species, but was later synonymized with *Plutellus* (Michaelsen 1900). Over the years it has become apparent that *Plutellus* and *Megascolides* were essentially polyphyletic waste bins for species which were not sufficiently different from each other morphologically to be distinguished by the few characters Michaelsen considered generically important. Jamieson (1972) pointed out that the North American species of *Plutellus* should be removed to *Argilophilus*, a suggestion more formally carried out by Gates (1977).

1-56670-053-1/95/$0.00+$.50
©1995 by CRC Press, Inc.

Macnab and McKey-Fender (1959a, b) recognized within the classic genera the existence of species groups which have more recently been accorded generic status (McKey-Fender and Fender 1982, McKey-Fender 1982, Fender and McKey-Fender 1990). In addition, we proposed (Fender and McKey-Fender 1990) to place all of the genera in a single tribe, the Argilophilini; for while it is clear that the fauna are morphologically diverse, the assemblage is obviously derived from one or at most two ancestors and the radiation has taken place in situ.

Each of the nine genera has a limited range within the region (see Figures 1 to 6 at the end of this chapter). For example, *Arctiostrotus* is found only from the moist forests of northwestern Oregon, through western Washington up to the Queen Charlotte Islands, whereas *Chetcodrilus* is in extreme southwestern Oregon and northern California along the coast. We are looking at this as evidence that radiation has been relatively recent. In fact, *Driloleirus macelfreshi* has a range entirely within an area which was inundated during the Pleistocene and buried under hundreds of feet of mud.

The nearest relatives of the Pacific Northwest Megascolecidae would appear to be southeast Asian (Fender and McKey-Fender 1990) Many of these (Gates 1972) have been described as *Plutellus* although they undoubtedly belong in their own genera. The Australian genera *Simsia* and *Graliophilus* (an anagram of *Argilophilus*) possess a great many similarities to our taxa as well, but in our opinion are somewhat more distant. *Diplocardia*, though its geographic range is close and it is in the same family, is very distant in its taxonomic relationships.

II.   Historical Biogeography

Three major factors have strongly affected the biogeography of earthworms in western North America: the pleistocene glaciation, the arid region to the east, and the human migration of the last four centuries.

Nearly the whole of Canada and the northern edge of the United States was covered by an ice sheet which undoubtedly eradicated earthworms over vast areas. In spite of this a relict fauna has been found in western Vancouver Island including four earthworm species, two of which are now widespread on the island, and one of which is found also on the Queen Charlotte Islands (McKey-Fender and Fender 1982).

Various European Lumbricidae have become the dominant species in most of the disturbed areas of the region. They apparently arrived through a variety of agencies (ship's ballast, potted plants, fishermen, and so forth) and once here have been able to disperse with remarkable speed. Fishermen, especially, have made sure that the lumbricids have had the chance to colonize the entire region. The remarkable fact is that they have not taken over in all areas, but only in certain habitats (see below).

An important question is why the native Megascolecidae have largely failed to reinvade areas free of glaciers for the last 10,000 years, while the Lumbricidae have managed to colonize these areas in a few centuries. Except in the extremely moist forest along the coast, the distribution of megascolecid earthworms stops very near the terminal moraines of the ice sheet. A possible answer lies in a preference of native Megascolecidae for fine-textured soils, which are largely absent in glaciated areas. Even in areas never affected by glaciation, the megascolecids are absent in sands (though sometimes present in the humus layers above) and do not compete well on clay-poor silts.

The arid region east of the Cascade Range and the Sierra Nevada has limited the distribution of the Argilophilini. Gates (1967) in his "Great American Desert" paper reported a large series of collections which turned up no endemic species, even in the moister parts of that region, the Great Basin of the western U.S. Further, the absence of Argilophilini in the eastern U.S. and the absence of *Diplocardia* in the Northwest argues for the permanence of the barrier. Still, there are a few populations in central Oregon (*Argilophilus hammondi*), eastern Washington (*Driloleirus americanus*), and central Idaho (*Arctiostrotus* spp.) that are now cut off from their relatives westward and show that there must have been a wider distribution in a moister past.

III. Ecology

Knowledge of the ecology of western North American earthworms is in a primitive state, with most of the information about them in taxonomic papers on individual species and in unpublished notes in our collections records. Only a few non-taxonomic papers have even touched on the existence of earthworms in natural conditions in the Pacific Northwest.

Dirks-Edmunds (1947) discusses relative abundance of earthworms in two forest communities, one in Oregon and one in Illinois, and notes that earthworms were abundant in the Cedar-Hemlock forest in Oregon (though the species involved were Megascolecidae, not Lumbricidae as stated in her paper).

Lotspeich et al. (1961), working on the relationships of vegetation and soil in Clallam County, Washington, mention earthworms only in their description of the soil profiles, but do talk about the absence of podzolisation, despite the acid pH and vegetation typical of podzolic soils elsewhere. The *Arctiostrotus* species involved is described in a taxonomic paper being submitted for publication this year.

By far the most detailed ecological work on western North American Megascolecidae involved another undescribed *Arctiostrotus* species (Spiers et al. 1984, 1986). This work is especially important since it is the first study to detail a whole range of ecological factors for a single western native earthworm species. The species was shown to be abundant at extremely low pH (2.9), while being present in a pH range of 2.6 to 6.2. It was not important in mixing of humus and soil, but was extremely important in the formation of a deep mor.

The casts themselves were shown "to account for up to 60% of the volume of the organic material in the humus profile" in many sites. The worms were feeding not only on fine organic debris, but also burrowing through and feeding on decayed logs. Plant nutrients were shown to be at higher availability in castings, as in similar studies with lumbricids. Calcium oxalate crystals in the earthworm ingesta were shown to disappear as the material proceeded through the intestine, supporting work by Cromack et al. (1977). There were strong indications as well of N-fixation within the earthworm gut or feces. The laboratory in which this work was done has since been disbanded.

This same species has been found in high numbers in old decaying logs on the western edge of the Olympic Peninsula, but only to a much lesser extent in non-woody forest litter or soil (Fender, unpubl. data). It is also described in the taxonomic paper mentioned above.

In western Oregon forests, Cromack et al. (1977) found megascolecid earthworms in *Hysterangium* (false truffle) mycelial mats. These mats are sites for accumulation of calcium oxalate in forest ecosystems and are important in overall calcium cycling. These and other earthworms have been found to harbour microorganisms which can digest the oxalate portion, thus freeing the calcium into a more available form.

While the formal studies of earthworm ecology in the western United States are rare, our collections for taxonomic work have been accompanied by notes on habitat and associated biota and it is on these observations that we base our overall view of the ecology of our western earthworms.

The native species differ ecologically to a remarkable extent from the adventive Lumbricidae. The Argilophilini appear to have a much higher tolerance than the Lumbricidae for soils with high clay content, low pH, and resinous, low nitrogen plant litter (whether this last is because of tolerance of polyphenols, tannins, high C:N ration, a combination of the above, or something else is unknown). Conversely, the Argilophilini seem unable to tolerate sandy soils which will still support lumbricid populations. It is not surprising then, that the Megascolecidae predominate in natural forest soils in most of the Pacific Northwest (especially where soils are derived from basic igneous rock and so have a high clay content) and are largely responsible for soil structure, mixing or nonmixing of organic matter into mineral layers, and are the base of many food chains involving insects and moles.

Soils derived from serpentine and related magnesium rich rocks sustain a rich Megascolecid fauna but are nearly free of Lumbricidae, even where the vegetation would appear to be of a type favoring lumbricids. While part of this effect may be because of the typically high levels of 2:1 clays in these soils, the earthworms from these sites also display a high tolerance for magnesium, being unaffected by the Epsom salt solution used in our laboratory to anaesthetize other earthworms (Fender and McKey-Fender 1990).

In western North America, while there are upwards of 100 native earthworms species, most species have limited ranges. In each location there are generally one to four native species, corresponding roughly to the following niches:

1. Upper litter and moss layers, feeders on fresh to partly decayed organic material, no mixing.
2. Lower humus layer, feeders on partly or well decayed organic material, no permanent burrows, some mixing.
3. Surface feeders with deep permanent burrows, generally large species.
4. Mineral soil, no permanent burrows, mixers.

The following list gives typical residents of a mixed coniferous forest near Corvallis, Oregon:

1. *Arctiostrotus perrieri*
2. *Toutellus toutellus*
3. *Driloleirus wellsi*
4. *Argilophilus panulirus*

Less than 32 km away in an open oak savanna, the species are:

1. Unrepresented
2. *Toutellus oregonensis*
3. *Driloleirus macelfreshi*
4. *Argilophilus garloughi*

A similar list of niches for Lumbricidae would be:

1. *Dendrodrilus rubidus*
2. *Octolasion tyrtaeum*
3. *Lumbricus terrestris*
4. *Aporrectodea turgida*

Areas with unusual conditions, such as areas of especially tight clay soils, tend to have only one or few species. In addition, other species which occur in mossy seeps and other specialized sites may add to the species richness of a forest, though not of the actual soil. Often, the species which occupy niche 1 will also be found in rotting wood and well oxygenated seeps.

Many Argilophilini display a reticulated distribution in the forest, being at much higher concentrations along game trails. This seems to be related to soil compaction, since human trails and other compacted areas show the same effect.

The saliva of many species has a flowery odor and is copiously produced in response to disturbance. This has never been examined for composition but we have hypothesized that it functions as a chemical defense.

IV.   Variation

Despite the cohesiveness of the Argilophilini from an evolutionary standpoint, the individual taxa show great diversity in their adaptations. Their habitats range from acid forest humus of the coastal forests to deep loess in dryland prairies of eastern Washington. Some of the smaller taxa are so delicate that they may not survive the trip from forest to laboratory. Their intestines will start to degrade within minutes of injury, in hours producing a jelly-like mass that is totally unusable for taxonomic purposes. The largest species, *Driloleirus macelfreshi*, can attain a length of nearly a meter and it is robust by any standards.

Similarly, *Arctiostrotus* species are generally very intolerant of heat and die at moderate room temperatures, whereas some large species are sufficiently heat tolerant to withstand being transported in a plastic bag in the back of an open pickup in early summer.

The typhlosole in the Argilophilini is extremely variable, though always present to some degree. In its most developed form it may be several times bifurcate in cross section; in its least developed form it is a barely present ridge on the intestine ceiling. Since the typhlosole increases the surface area of the intestine it would seem this would have bearing on the ecology of the species involved, but that has not been investigated.

Most of the Argilophilini have no pigment, though even some normally unpigmented species may show a flush of orange at sexual maturity (e.g. *Driloleirus wellsi*). A few species (i.e. *Arctiostrotus perrieri*, *A. altmani*, and an undescribed *Toutellus* sp.) have a reddish brown pigmentation similar to *Lumbricus rubellus*.

Activity levels also vary tremendously in these taxa. *Arctiostrotus perrieri*, when disturbed, will flip back and forth, jumping several centimeters off the ground. *Toutellus oregonensis* will shorten into a stiff phallic shape and not move. Most of the species are never seen on the surface, but the very large *Driloleirus wellsi* will occasionally be seen by joggers along wet trails on misty days. *Driloleirus* species generally form permanent burrows at least 4.5 meters deep (Smith 1897) and can move very rapidly to escape from a shovel.

While the above remarks all pertain to the native Megascolecidae, it is well to note that the limicolous genus *Sparganophilus* is also widespread in western Oregon and California and is very important in stream margins and some bogs. Along river banks, *Sparganophilus* species will feed in silt and buried leaf mats sufficiently low in oxygen to have a characteristic odor. There have been no new descriptions of western *Sparganophilus* since Eisen (1896), but there appear to be several new species in the Pacific Northwest.

V.   Research Imperatives

A.   Ecological Studies

There are many opportunities for ecological research on earthworms in the western United States. I would especially like to see an examination of the saliva of some species as a possible chemical defense and an in-depth look at the species on serpentinitic soils.

B.   Taxonomic Needs

The primary need, however, is for species descriptions of all, or at the very least most, of the native species. With such a large proportion of the fauna undescribed, it is almost guaranteed that field researchers will be unable to identify the taxa in their sites. Further, the literature is already too scattered to be useful for ecologists. What is really necessary is a monograph on the western species, getting all of the information into one place and describing all of the 100+ species in our collection. Unfortunately, such a study would take about five years of full-time work and would result in a very large manuscript, too large for publication in a journal. Once done, however, it would be the basis for research for many decades.

I do not see this as a crusade to identify a few unimportant species before they are wiped out by the invasive Lumbricidae. Only a very few of the Argilophilini (e.g. *Driloleirus macelfreshi*) can be said to be endangered, since lumbricids do so poorly in the forests of our region. The real issue is that they are the primary developers of forest soils throughout much of our region and are likely to remain important in those soils into the foreseeable future.

References

Cromack, K., Jr., P. Sollins, R.L Todd, R. Fogel, A.W. Todd, W.M. Fender, M.E. Crossley, and D.A. Crossley Jr. 1977. The role of oxalic acid and bicarbonate in calcium cycling by fungi and bacteria: Some possible implications for soil animals. In: Soil organisms as components of ecosystems. *Ecol. Bull. (Stockholm)* 25:246–252.

Dirks-Edmunds, J.C. 1947. A comparison of biotic communities of the cedar-hemlock and oak-hickory associations. *Ecological Monographs* 17:235–260.

Eisen, G. 1896. Pacific coast Oligochaeta. II. Mem. *Calif. Acad. Sci.* 2:123–199

Fender, W.M. and D. McKey-Fender. 1990. Oligochaeta: Megascolecidae and other earthworms from Western North America. p. 357–378. In: Dindal, D. (ed.) *Soil biology guide*. Wiley and Sons, New York.

Gates, G.E. 1967. On the earthworm fauna of the Great American Desert and adjacent areas. *Great Basin Nat.* 27:142–176.

Gates, G.E. 1972. Burmese earthworms. An introduction to the systematics and biology of megadrile oligochaetes with special reference to southeast Asia. *Trans. Am. Phil. Soc.* 62:1–326.

Gates, G.E. 1977. On the correct generic name for some West Coast native earthworms, with aids for a study of the genus. *Megadrilogica* 3:54–60.

Jamieson, B.G.M. 1972. The Australian earthworm genus *Spenceriella* and the description of two new genera (Oligochaeta: Megascolecidae). *Mem. Nat. Mus. Vic.* 33:73–88.

Lotspeich, F.B., J.B. Secor, R. Okazaki, and H.W. Smith. 1961. Vegetation as a soil-forming factor on the Quillayute Physiographic Unit in Western Clallam County, Washington. *Ecology* 42:53–68.

Macnab, J. A. and D. McKey-Fender. 1959a. A new species of *Plutellus* from western Oregon (Oligochaeta: Megascolecidae). *Northwest Sci.* 33:69–75.

Macnab, J.A. and D. McKey-Fender. 1959b. Two new species of *Plutellus* from Coos County, southwestern Oregon (Oligochaeta: Megascolecidae). *Northwest Sci.* 33:157–159.

McKey-Fender, D. 1982. Dr. Kincaid's other worm. *Megadrilogica* 4:86–88.

McKey-Fender, D. and W.M. Fender. 1982. *Arctiostrotus* (gen. nov.). I. The identity of *Plutellus perrieri* Benham 1892 and its distribution in relation to glacial refugia. *Megadrilogica* 4:81–86.

Michaelsen, W. 1900. Oligochaeta. In: *Das Tierreich*. Lief 10. Verlag von Friedländer und Sohn, Berlin.

Reynolds, J.W. 1980. The earthworm family Sparganophilidae (Annelida, Oligochaeta) in North America. *Megadrilogica* 3:189–204.

Smith, F. 1987. Upon an undescribed species of *Megascolides* from the United States. *Am. Nat.* 31:202–204.

Spiers, G.A., D. Gagnon, G.E. Nason, E.C. Packee, and J.D. Lousier. 1986. Effects and importance of indigenous earthworms on decomposition and nutrient cycling in coastal forest ecosystems. *Can J. For. Res.* 16:983–989.

Spiers, G.A., G.E. Nason, J.D. Lousier, V. Marshall, E.C. Packee, and D. Gagnon. 1984. Endemic earthworms and nutrient cycling in Coastal Wet Western Hemlock Biogeoclimatic Subzone, Vancouver Island. In: Stone, E. (ed.) Forest soils and treatment impacts: Proceedings of the 6th North American Forest Soil Conference. University of Tennessee. Knoxville. *Pub. Soc. Am. For.* 84-10. p. 447.

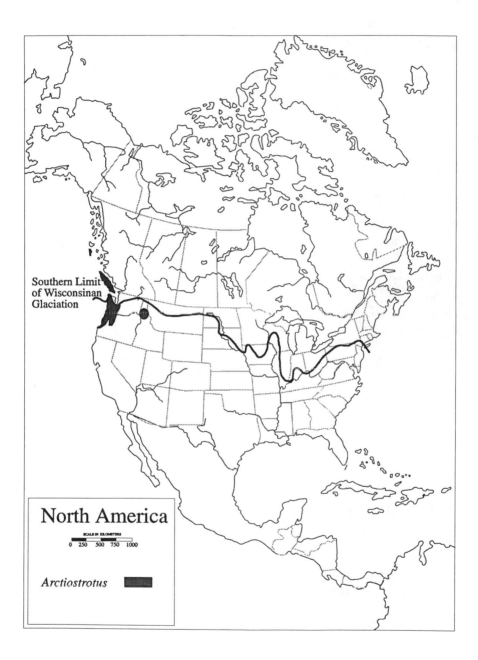

Figure 1. The approximate North American distribution of *Arctiostrotus*.

Figure 2. The approximate North American distribution of *Argilophilus* and *Kincaidodrilus*.

Figure 3. The approximate North American distribution of *Drilochaera* and *Chetcodrilus*.

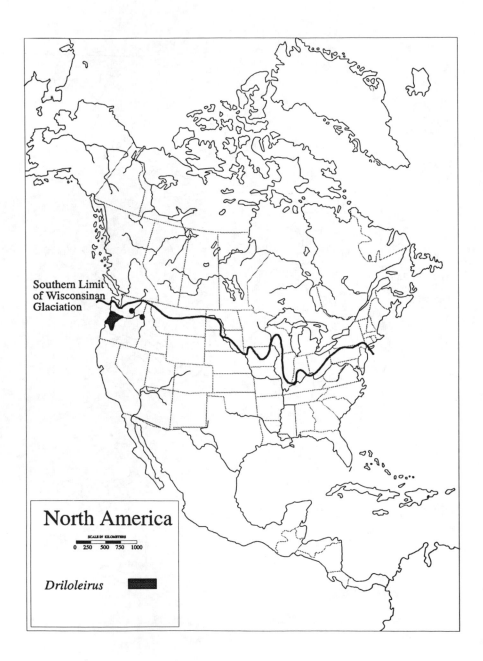

Figure 4. The approximate North American distribution of *Driloleirus*.

Figure 5. The approximate North American distribution of *Macnabodrilus* and *Nephrallaxis*.

Figure 6. The approximate North American distribution of *Toutellus*.

Native Earthworms of the North Neotropical Region: Current Status and Controversies

C. Fragoso, S.W. James, and S. Borges

I. Introduction

Notwithstanding Darwin's publication of a book on earthworms more than one hundred years ago (1881), there still are large gaps in the knowledge of earthworm ecology, distribution, and species diversity. This is particularly true for Mexico, Central America, and the Caribbean, where ecological and taxonomical studies are scarce. In the present chapter the current status of earthworms in this region is presented, with a particular emphasis on taxonomy, biogeography and the impact of disturbance on native species. A historical perspective is also given, along with a review of the biogeographical controversies and the kind of research that is urgently needed.

A. The North Neotropical Region (NNR)

The neotropical region has been defined in biogeographic terms on the basis of animal and plant distributions (Pielou 1979). It includes South and Central

1-56670-053-1/95/$0.00 + $.50

America, the Caribbean islands, and part of Mexico (the northern limit being the Sonoran and Chihuahuan deserts).

The north neotropical region excludes South America but includes Central America, the Caribbean Islands, and the southern part of Mexico. Its extent may be delimited in both continental and insular terms, to the north by the great Mexican deserts and Cuba, to the south by the Darien forest on the Panama-Colombia border and the island of Grenada.

This region can be further divided into two important subregions: the northern which includes Mexico, the Greater Antilles, Guatemala, Honduras, Nicaragua, El Salvador, and the northern half of Costa Rica; and the southern which covers the southern half of Costa Rica, Panama, and the Lesser Antilles.

The high environmental heterogeneity of this region is manifested by its diversity of climates, soils and orographic landscapes. As a result it is possible to find many different kinds of vegetation, from warm tropical savannas and tropical rain and deciduous forests to cool temperate cloud forests and pine and oak forests. Obviously this heterogeneity promotes earthworm diversity.

B.   Historical Perspective

The first records and descriptions of earthworms from this region were made by Rosa, Beddard, Benham, Eisen, Michaelsen, and Cognetti at the end of the 19th century and the beginning of the 20th century (Table 1). After this period of intensive research, studies on earthworm taxonomy drastically declined. In a period of more than 30 years (1923–1956) only three papers appeared, primarily concerned with taxonomic descriptions of earthworms from Honduras and the Caribbean islands (Michaelsen 1935), the Mexican Yucatán peninsula (Pickford 1938), and the first checklist of the region (Gates 1942). Since 1957, mainly due to the work of G.E. Gates, the research on earthworm taxonomy again increased (Table 1). Gates (1982) summarized the knowledge of North American earthworms in a paper that marked the end of his scientific career. In more recent years several taxonomic papers have appeared, mainly from Jamaica and Puerto Rico (Sims 1987, Borges 1988, 1992, in press a, Borges and Moreno 1989, 1990b, 1991, Csuzdi and Zicsi 1991, James 1991), Costa Rica (Righi and Fraile 1987) and Mexico (Fragoso 1988, 1989, 1991, 1993, James 1990, 1993, Zicsi and Csuzdi 1991, Csuzdi and Zicsi 1991, and Fragoso and Rojas 1994).

Table 1. Summary of taxonomic research on earthworms in the north neotropical region.

Authors	Period	No. of Papers	Described Species
Rosa	1891	1	1
Beddard	1890, 1893	2	2
Benham	1896	1	1
Eisen	1893–1900	4	17
Michaelsen	1890–1935	11	18
Cognetti	1904–1908	6	7
Pickford	1938	1	2
Graff	1957	1	8
Gates	1942–1982	18	19
Righi	1972	1	3
Righi and Fraile	1987	1	6
Fragoso	1988, 1991	2	5
Fragoso and Rojas	1994	1	5
Borges	1994	1	1
Borges and Moreno	1989–1992	3	12
Csuzdi and Zicsi	1991	1	3
Sims	1987	1	2
James	1990–1994	3	9
Total	1890–1994	59	121

See Gates (1942, 1982) for references not quoted in the text.

II.   Current Status

A.   Taxonomy and Ecology of Earthworms from the NNR

In this section the current status of research in the countries that make up the NNR will be presented. A summary of this information is provided in Table 2. The complete list of species in this region is shown in Appendix 1. Figures 4 to 15 at the end of this chapter show the distribution of the genera present in this region.

C. Fragoso, S.W. James, and S. Borges

Table 2. Number of species and genera of native earthworms in Mexico, Central America, and the Greater Antilles.

| Country | Megascolecidae | | | | Ocnerodrilidae | | Glossoscolecidae | | Total | |
| | Acanthodrilini | | Dichogastrini | | | | | | | |
	Species	Genera	Species	Genera	Species	Genera	Species	Genera	Species	Genera
Mexico	28	9	18	4	1	1	3	3*	50	17
Guatemala	4	2	9	2	1	1	1	1	15	6
El Salvador	2	2	6	2					8	4
Honduras			2	2					2	2
Costa Rica	1	1	11	2	2	2	6	2	19	6
Cuba			1	1			2	1	4	3
Jamaica	1	1	9	2			1	1	10	3
Haiti			6	1	4	1	1	1	8	3
Dominican Re- public			1	1					5	2
Puerto Rico			7	3			11	3	18	6

The table only includes described species, excepting Mexican glossoscolecids (*). Although two of these genera have not yet been described, they are included here because of their biogeographical importance.

Central to this chapter is the distinction between native and exotic species. Native species are found naturally within a given area, whereas exotic ones have invaded several sites through human dispersal. Endemic species are natives that presumably evolved in the sites where they are found.

1.   Mexico
a.   Taxonomy

With a surface area of 1,958,201 km², Mexico ranks as the 14th largest country in the world and the largest in the NNR. Its complex topography translates into a large variety of environmental conditions, which in turn produce more than 10 major vegetation types. At present, more than 50% of the total area of Mexico has been disturbed (Flores and Gerez 1988). As will be seen later, the conversion of natural vegetation into managed lands has drastically affected native species.

The earthworm fauna of Mexico consists of 88 described species which, in turn, are divided into 48 native and 40 exotic species. Recently, Fragoso (unpublished data), Fragoso and Rojas (unpublished data), and James (unpublished data) have found at least 17 new undescribed species, giving a conservative estimate of 105 species. The native earthworm fauna of Mexico falls into the families Megascolecidae (Tribes Acanthodrilini and Dichogastrini), Ocnerodrilidae, and Glossoscolecidae. Table 2 clearly shows that the Megascolecidae predominate, accounting for more than 90% of the total number of species.

Several species of *Dichogaster* and one species of *Sparganophilus* (presumably *eiseni*) were excluded from the list of natives, because of their questionable origins.

Acanthodrilini: This tribe includes 26 genera distributed worldwide, except in the Paleartic region. Fragoso (1993) and Fragoso and Rojas (in press) divide this tribe into eight groups, on the basis of male genitalia, nephridial vesicles, calciferous glands, number of setae, and number of gizzards.

In Mexico the acanthodrilinae are represented by the genera *Diplotrema*, *Balanteodrilus*, *Lavellodrilus*, *Zapotecia*, *Diplocardia*, *Mayadrilus*, *Kaxdrilus*, *Protozapotecia*, and *Larsonidrilus*.

Balanteodrilus is restricted to eastern and southeastern Mexico and includes *Balanteodrilus pearsei* and two more undescribed species (Fragoso and Rojas unpublished data). The majority of the records correspond to the former species (Pickford 1938, Gates 1977a, Fragoso and Lavelle 1987, Fragoso 1990, 1992). *Lavellodrilus*, another endemic earthworm genus from eastern and southeastern Mexico, includes six species. *Diplocardia* is primarily a nearctic genus with only three endemic species in northern and central Mexico. *Zapotecia* is distributed throughout the central mountain region (Trans-Mexican volcanic belt—TVB) and in northeast Mexico, and includes two species and one more undescribed (Fragoso 1993). *Protozapotecia* is a recently described genus situated between

Diplocardia and *Zapotecia* with a distribution restricted to central and east Mexico; it includes two species (James 1993).

Diplotrema is a very complex genus whose history has been discussed by Fragoso (1988) and James (1990); it is only briefly summarized here. Jamieson (1971), after stating that *Eodrilus* was no longer a valid name, separated all the species of this genus into *Diplotrema* and *Notiodrilus*. In the former he placed only the Australian species, and in the latter all the non-Australian species (48 species). A preliminary revision of these two genera (Fragoso and Rojas 1994) places the species of Mexico, the Caribbean, and Central America into *Diplotrema* and *Kaxdrilus*, newly described genus. In Mexico, *Diplotrema* includes 7 species distributed in the eastern and southeastern region with only one record in the western region (*D. vasliti*). *Kaxdrilus* is mainly confined to natural forests south of the Tehuantepec Isthmus (TI) and is represented by three species in Mexico (Fragoso and Rojas 1994).

Two additional new genera from the Yucatan peninsula (*Mayadrilus*) and from central-east Mexico (*Larsonidrilus*) are described by Fragoso and Rojas (1994) and James (1993), each one containing two species.

Dichogastrini: This tribe contains 40 genera, with a geographic distribution very similar to that of the acanthodrilines. In Mexico, five genera have been recorded: *Eutrigaster*, *Dichogaster*, *Ramiellona*, *Zapatadrilus,* and a new undescribed genus (Fragoso and Rojas in prep.).

The American species of *Dichogaster* were separated by Sims (1987) into the genera *Dichogaster* (with penial setae and with two gizzards in vi and vii) and the resurrected *Eutrigaster* (without penial setae and with two gizzards in vi and vii and a muscular proventriculus in v). Later, Csuzdi and Zicsi (1991) restricted *Dichogaster* to a small group of species with the first dorsal pore before segment x; in *Eutrigaster* they left the other species characterized by a first dorsal pore in x (or afterwards) and the muscular proventriculus in v. The last genus was in turn subdivided into two subgenera on the basis of the presence (*E. (Graffia)*) or absence (*E. (Eutrigaster)*) of penial setae. In Mexico the genus *Eutrigaster* includes five species, and is mainly found in the central (TVB) and southern mountain systems. The species *E. paessleri,* was synonimized with *E. (G.) viridis* by Csuzdi and Zicsi (1991), and thus is not included here. *Dichogaster* includes three species from west Mexico that could be exotic. However, they are tentatively considered as natives in the hope that a future monographic study will confirm or deny this status.

The genus *Ramiellona* presents the bulk of species in Central America. In Mexico it includes six species, five already described and one more in preparation (Fragoso 1993). Their distribution area is south of the TI.

A new undescribed genus similar to *Ramiellona* was separated by Fragoso (1993) after he documented that the distinctive characteristics of *R. mexicana* (Gates 1962a) and *R. willsoni* (Righi 1972b) were also present in several new species found in southeastern Mexico. At present this new genus includes 13 species (two described) restricted to Mexico, their distribution being well demarcated between TI and TVB.

Zapatadrilus was proposed by James (1991) for the species of *Trigaster* with two gizzards. In Mexico five species have been reported, with two more species in the process of description (Fragoso 1993). Their distribution occurs north to the TVB, either in the mountains or plains.

Ocnerodrilidae: Only one native genus of this family has been found in Mexico. *Phoenicodrilus*, described by Eisen (1895) and resurrected by Gates (1977b), contains the peregrine species *P. taste* (distributed in most of the country) and a new one, still undescribed (Fragoso 1993).

Glossoscolecidae: Fragoso (1993) has observed three native species in this family from eastern and southeastern Mexico. Two of these species each represent a new genus. The other species is *Pontoscolex cynthiae*, a recently described species from Puerto Rico (Borges and Moreno 1990b). The occurrence of *P. cynthiae* in swamp soils of Tabasco state (Fragoso 1993) is relevant in terms of the biogeographic relations between Mexico and the Caribbean.

Before this report, it was assumed that the northern continental limit of the family was Costa Rica. These new findings, however, require a change in this limit and a modification of current biogeographic interpretations.

b.   Ecology

All of the ecological studies in Mexico have been centered in the eastern and south-eastern region and include the study of earthworm communities of tropical rain forests (Lavelle and Kohlmann 1984, Fragoso 1985, 1992, Fragoso and Lavelle 1987, 1992), pastures and tropical subdeciduous forests (Lavelle et al. 1981), and altitudinal studies in tropical subdeciduous, cloud, and coniferous forests in east Mexico (Fragoso 1989).

Fragoso (1993) summarized this information and, with additional data, arrived at the following general conclusions:

1. The earthworm communities of eastern and southeastern Mexico are dominated by endogeic-geophagous species. This pattern can be explained by phylogenetic constraints (dominance of primitive Megascolecidae) and by the high nutrient enrichment of soils in this region.
2. The structure of these communities is influenced by soil moisture (acting upon species richness and ecological categories), temperature (ecological categories), and soil nutrient enrichment (species richness and abundance). A summary of this information is shown in Table 3.
3. Only in a few situations is competition the best explanation for certain morphological and ecological patterns.

2.   Central America (excluding Panama)

In the majority of Central American countries, research on earthworms has been scarce. Moreover, in two countries (Belize and Nicaragua) there are no records of native species. (Note: While this paper was in press, Reynolds and Righi (1994) described a new species from Belize.)

   Guatemala. Eleven of the fifteen species of Guatemala were described by Eisen (1896, 1900) and Michaelsen (1911, 1912). Gates (1962a) reviewed the collection made by Eisen in 1900, and described several more species. No other surveys of earthworms have been carried out since then.

   The family Megascolecidae is the best represented with 13 species distributed into Dichogastrini (nine species) and Acanthodrilinae (four species). The former includes the genera *Eutrigaster* (two species) and *Ramiellona* (seven species). The Acanthodrilini includes one species of the genus Diplotrema, and three species of the genus *Kaxdrilus* (Fragoso and Rojas 1994).

   The families Ocnerodrilidae and Glossoscolecidae are represented by *Ocnerodrilus tuberculatus* (Eisen 1900) and *Pontoscolex lilljeborgi* (Eisen 1896), respectively.

   El Salvador. Knowledge of the earthworm fauna of this country is entirely based on the paper of Graff (1957), who recorded two species of Acanthodrilinae and six species of Dichogastrini. The former species belong to the genera *Diplotrema* and *Kaxdrilus* (Fragoso and Rojas 1994), whereas the dichogastrini are placed in the genera *Eutrigaster* (three species) and *Ramiellona* (three species).

   Honduras. Only two dichogastrini species have been reported in this country: *Ramiellona stadelmanni* (Michaelsen 1935) and *E. (G.) sporadonephra* (Righi 1972b).

Table 3. Summary of the effect of temperature (T), moisture (M), and soil nutrients (SN) on the structure of earthworm communities in the east Mexican tropics (modified from Fragoso, 1993). EN=endogeics, EP=epigeics, ↑=high, ↓=low.

	M+ T+	M+ T−	M− T+	M− T−
SN+	EN>EP ↑↑abundance ↑diversity *Tropical rain forests*	EP=EN ↓ abundance ↓ diversity *Cloud forests*	EN>EP ↓ abundance ↓ diversity *Tropical subdecidous forests*	EN=EP ↑ abundance ↓ diversity *Pine-Oak forests*
SN−	EN>EP ↑ abundance ↓ diversity *Riparian environments*	not found in this study	not found in this study	not found in this study

Costa Rica. With 19 native species, Costa Rica is the best surveyed country of Central America. The fauna is dominated by the family Megascolecidae, tribe Dichogastrini (11 species), with no records of Acanthodrilini. The dichogastrine species belong to the genera *Eutrigaster* (seven species) and *Dichogaster* (four species).

The family Glossoscolecidae is represented by six species in the genera *Andiodrilus* (two species) and *Glossodrilus* (four species).

Finally, the Ocnerodrilidae are represented by *Ocnerodrilus alox* (Righi and Fraile 1987) and *Nematogenia panamaensis* (Cognetti 1904b).

Few ecological studies have been made in this country, e.g., Picado (1913) who studied bromeliad fauna and Atkin and Proctor (1988) who studied density and biomass of litter and soil invertebrates in altitudinal gradients. Only in the agronomical study of Fraile (1989) were earthworms the central theme of investigation.

Greater Antilles. The oligochaete fauna of the Greater Antilles (GAs) has been poorly studied. Most of the earthworm records for the islands date from the late 1800's and early 1900's. Still, those contributions have been scarce and incidental to other studies, even though other groups of animals have been surveyed extensively.

Cuba. The largest, most ecologically diverse and least-surveyed island of the archipelago only has four native species reported: one Acanthodrilinae (*Diplotrema ulrici*), one Dichogastrini (*Zapatadrilus cavernicolus*) and two glossoscolecids (genus *Onychochaeta*). Recently C. Rodríguez (personal communication) has found three new species of *Zapatadrilus*, and several dichogastrines (*Eutrigaster (Graffia)* spp.), ocnerodrilids and glossocolecids. Excepting the recent paper of Rodríguez (1993), no systematic studies concerning earthworms have been carried out in Cuba. Scientists on the island, however, have been examining other aspects of these organisms. Recent publications entail morphological studies and the distribution of *Polypheretima elongata* (Rodríguez and Reines 1986, 1989), several aspects of the biology of *Eudrilus eugeniae* (Rodríguez et al. 1986, 1987, 1988, Rodríguez 1991), and the use of electrophoresis and esterases as taxonomic tools (Rodríguez and Gonzalez 1986). In addition, vermiculture has been carried out successfully on the island for several years. In fact, by 1990 they expected to obtain 60,000 tons of humus (Campa et al. 1990) from their earthworm farms.

Jamaica. From a taxonomic perspective, the oligochaete fauna of Jamaica has been moderately well studied. There are now nine native species of Dichogastrini and one of Glossoscolecidae (*Diachaeta thomasi*). In the former group the species belong to the genera *Dichogaster* (one species) and *Eutrigaster* (eight species). James (unpublished data) has found other new species of *Eutrigaster*.

Laessle (1961) reported green earthworms living between the leaves of bromeliads. The worms probably belong to *Dichogaster*.

Hispaniola. The oligochaete fauna of Hispaniola has been completely disregarded since the 1930's. Only eight native species have been reported from Haiti and five from the Dominican Republic. In the former country the species are distributed into the Acanthodrilinae (*Zapotecia keiteli*), Dichogastrini (*Eutrigaster* (*Graffia*), six species) and Glossoscolecidae (*Onychochaeta windlei*), whereas in the Dominican Republic they are placed in the Dichogastrini (*E. (G) godeffroyi*) and Ocnerodrilidae (four species of *Temanonegia*, Gates 1979).

Puerto Rico. Puerto Rico is the smallest of the Greater Antilles and is the island whose oligochaete fauna has been the most studied. The first records for the island were two peregrine species mentioned by Michaelsen in 1902. More than fifty years elapsed before Gates (1954) described the first new species from the island. Then Gates (1970a) described the genus *Estherella*, the first native glossoscolecid genus of the Antilles. Recent investigations by Borges (1988, 1992, in press a), Borges and Moreno (1989, 1990a, 1990b, 1991, 1992, 1994) and James (1991) have greatly expanded the knowledge of earthworms on the island. Eighteen native species are now known from Puerto Rico, 11 placed in Glossoscolecidae (seven species of *Estherella*, three species of *Pontoscolex*, and *Onychochaeta borincana*) and seven in Dichogastrini (three species of *Borgesia*, three species of *Neotrigaster*, and *Trigaster longissimus*).

With the alpha taxonomy practically completed, other investigations have been done on the ecology of these organisms in Puerto Rico. One was carried out in the Laguna Cartagena Wildlife Refuge (Alfaro 1992) and the other in the Nipe soils (Oxisols) of the Maricao State Forest (Hubers 1993). In both studies the earthworm abundance and diversity were found to be lower than those reported in other Neotropical regions. Furthermore, the International Institute of Tropical Forestry (USDA) and the University of Puerto Rico are presently sponsoring a two-year project on the diversity and ecology of the earthworms at El Verde (formerly studied by Moore and Burns 1970) and Bisley Watersheds of the Luquillo Experimental Forest (Borges, in press b).

3.   The Lesser Antilles South of St. Kitts

Previous reports (summarized in Gates, 1982 and Righi, 1972a) listed a few species of Glossoscolecidae and one peregrine *Dichogaster* (Table 4; species marked with asterisks). The Glossoscolecidae include species native to Trinidad and two others of uncertain status. In September and October of 1991 earthworm specimens were collected from the larger of the Lesser Antilles (LAs), except Martinique and islands north of Guadeloupe, but including Trinidad and Tobago (James unpub.). The latter, though not Antillean, were included because they are on a potential dispersal pathway northward from South America. In 1993 more material was collected on Martinique, St. Kitts, Nevis, and Montserrat and we learned of an earlier collection from Martinique, now being studied by Moreno (Moreno pers. comm.).

The results of these surveys (Table 4; entries without astericks) support the hypothesis that few or no native species exist on Grenada, Barbados, St. Vincent, St. Lucia, St. Kitts, Nevis, and Montserrat. Trinidad and Tobago are geologically parts of South America (Pindell and Barrett 1990, Ladd et al. 1990), and have several native species known only from those islands or from

Table 4. Earthworms of the Lesser Antilles plus Trinidad, Tobago, and northern Venezuela. *Dichogaster* species are indigenous (I) or uncertain (U). Entries followed by an asterisk (*) are from previously published data (see text); the rest are from unpublished surveys. Numbers indicate the number of species within the genus on the island. Numbers followed by a small letter "n" are numbers of undescribed species, there being no described species of that genus from that location; a plus sign (+) indicates that additional new species are in the authors' collections. The right three columns indicate presence of three peregrine taxa. This list is not complete.

	Genus	No. of Species	*Pontoscolex corethrurus*	*Amynthas* spp.	*Eudrilus eugeniae*
St. Kitts	*Dichogaster* (U)	1	Y	Y	Y
Nevis	*Dichogaster* (U)	1	Y	Y	Y
Montserrat	*Dichogaster* (U)	1	Y	Y	
Guadeloupe	*Dichogaster* (I)	9	Y	Y	
	Neogaster	1n			
Dominica	*Dichogaster* (I)	5n	Y		
	Dichogaster (U)	1*			
	Andioscolex	1n			
Martinique	*Dichogaster* (I)	8n	Y	Y	
	Andioscolex	1n			
St. Lucia	*Dichogaster* (U)	2	Y	Y	
St Vincent	*Dichogaster* (U)	2	Y	Y	
Barbados	*Diacheta*	1	Y	Y	
Grenada	*Dichogaster* (U)	1	Y	Y	
Tobago	*Dichogaster* (U)	1	Y	Y	
	Onychochaeta	1			
	Rhinodrilus	1			
Trinidad	*Dichogaster* (U)	2	Y		
	Diacheta	1*			
	Onychochaeta	1			
	Rhinodrilus	1+*			
Venezuela	*Dichogaster* (I)	many	Y		
	Diacheta	1*			
	Onychochaeta	2*			
	Rhinodrilus	many			

Venezuela. Guadeloupe is composed to two islands. The eastern one comes from the central part of the old volcanic arc that gave origin to the GAs. The western part originated "in situ" and is part of a much younger volcanic arc that goes from Saba in the north to Granada in the south. Dominica and Martinique are part of this younger arc, but being so close to Guadaloupe, and because of the prevailing currents, indigenous earthworms from Guadaloupe could reach them. Only Dominica, Martinique, and Guadeloupe have what appear to be clearly indigenous earthworm species. The many very small *Dichogaster* spp. found in logs on the other Lesser Antille Islands surveyed have not yet been identified or thoroughly compared among islands. Their status is therefore still unresolved. However, such worms were only found in decaying wood and never in soil. No related taxa were found in soil, only peregrine species. On the islands with endemic species, the native earthworms were present in soils, logs, and epiphytes, and were easy to find.

High elevation areas remote from cultivation and from peregrine species are often small and difficult to reach. Where it was possible to collect from such places, only on Dominica, Martinique, Guadeloupe, St. Kitts, Nevis, and Montserrat could earthworms be found, even though soil moisture appeared adequate in all sites. The last three islands had exotic species (*Amynthas* spp., *Pontoscolex corethrurus*, *Eudrilus eugeniae* and *Neogaster divergeus*). The greater litter depth and F-layer development in several wormless sites (Grenada, St. Lucia) were similar to what has been observed on worm-free soils in temperate zone forests.

It is likely, therefore, that Dominica, Martinique, and Guadeloupe still have abundant populations of many undescribed earthworm species living in mountainous forested areas, including on the slopes of active volcanoes and in the most remote, undisturbed montane forests. Oddly, none have been found in the low elevation forest remnants, where only the peregrine *Pontoscolex corethrurus* occurs. The new species found so far belong to the megascolecid genus *Dichogaster* (s.l.) and to the glossoscolecid genus *Andioscolex*. No independently verifiable explanation for the presence of these earthworms is presently available.

So far it appears that the Lesser Antilles (with three significant exceptions) do not have native earthworms, except possibly some small epigeic species. In either case there is little evidence to support the hypothesis that earthworms have dispersed northward from South America, as have many other elements of the Antillean biota (Humphries and Parenti 1986), or eastward from Central America and Mexico into the LAs via the GAs.

B.   Earthworm Biogeography in the Northern Neotropic Region

1.   Mexico and Central America

On the basis of native megascolecid genera distributions it is possible to make a preliminary division of Central America and eastern and southeastern Mexico into three biogeographical regions (Fragoso 1993): the northern region (north to the TVB) characterized by the presence of *Zapatadrilus, Diplocardia, Zapotecia*, and *Protozapotecia*; the central region (between the TVB and the TI), dominated by a new dichogastrine genus related to *Ramiellona* (Fragoso 1993), and the southern region (south of TI, including El Salvador and Honduras) defined by the genera *Ramiellona* (mainly in highlands), *Lavellodrilus, Kaxdrilus*, and *Mayadrilus*.

The genera *Diplotrema, Eutrigaster* and *Balanteodrilus* show a particular distribution. *Diplotrema* is found in lowlands of all three regions. The genus *Eutrigaster* is found primarily in the highlands (over 1000 m) of the three regions including Costa Rica; if this genus occurs at altitudes below 1000 m, it is always inside bromeliads and/or decaying logs (e.g., *E. sporadonephra* in bromeliads of tropical forests of southeastern Mexico and Costa Rica; Fragoso and Lavelle 1987, Picado 1913). This pattern of geographic distribution suggests that *Eutrigaster* recently invaded Mexico and north Central America, through a real corridor when South and North America entered into contact. Finally *Balanteodrilus* is distributed in the three regions but always in the lowlands; allopatric speciation within this genus, however, occurred in a south-north direction (Fragoso and Rojas, unpublished data).

The glossoscolecid genera are limited to southern lowlands, being particularly well adapted to pasture habitats. Their distribution and habits suggest a recent invasion (in geologic terms) probably when the first contacts between North America and South America occurred (filter bridges at the end of the Cretaceous, Stehli and Webb 1985).

Former patterns indicate that within the Acanthodrilini and Dichogastrini, the primitive genera (*Diplotrema, Lavellodrilus, Balanteodrilus,* the new, undescribed, *Ramiellona*-related genus) are always located to the south or in lowlands, whereas the derived genera (*Diplocardia, Zapotecia, Protozapotecia, Kaxdrilus, Zapatadrilus, Ramiellona*) are located to the north or in highlands.

Although a great deal of research in historical geology and tectonic movements is still necessary, there appears to be some congruence between earthworm distribution and historic events. The following interpretation is based upon the works of Maldonado-Koerdell (1964), Ferrusquia (1993), and Donnelly (1985, 1988), and assumes that acanthodriline and dichogastrine ancestors were present in Pangea nearly 230 million years ago.

During the Triassic most of central Mexico was covered by epicontinental seas, and all the actual Gulf Coastal Plain was submerged. After the break-up of Pangea (200 million years ago), acanthodriline and dichogastrine species were separated in southern (Gondwana) and northern (Laurasia) stocks. In the former remained the ancestors of African, Australian, and South American megascolecids, whereas the latter gave rise to all the North American acanthodriline fauna, and some elements of dichogastrines and ocnerodrilids (as mentioned previously, *Eutrigaster* represents a recent invasion). During more than 100 million years, Mexico and part of Central America were intermittently covered by epicontinetal seas. At the beginning of the Cretaceous the sea divided Mexico at the level of the TI (López-Ramos 1981), or slightly north of the TVB (Maldonado-Koerdell 1964). It is quite probable that this event corresponded to the separation of the Chortis block from the Mexican and Maya blocks (Donnelly 1988), and thus represents the vicariant event that separated the north (the ancestors of *Diplocardia, Zapotecia, Protozapotecia, Zapatadrilus*) and south (the ancestors of *Ramiellona, Kaxdrilus,* and *Balanteodrilus*) earthworm faunas. The Chortis block rejoined Mexico at the end of the Cretacious (65 million years ago), when the Yucatan peninsula began to emerge. Subsequently the Laramidique orogenesis during the Cenozoic elevated Central America and the north and central parts of Mexico. At the end of the Eocene and during the Miocene the following systems were conformed: the East Sierra Madre, the Oaxaca and Chiapas mountain systems, the West Sierra Madre, and the Central America mountain systems. The formation of these mountain systems probably represented the vicariant events that promoted further differentiation between and within the dichogastrine and acanthodriline genera. In spite of the fact that the TVB now clearly separates north and central regions, their late emergence at the beginning of the Pliocene (5 million years ago), precludes it as the causal event of generic differentiation.

The former interpretation is still speculative, but offers interesting aspects to be solved in the future. Clarification of phylogenetic relationships in the Mexican and Central American earthworm fauna will be a critical step to achieve this goal.

2.   Lesser Antilles

When considering the biogeography of a group of islands such as the LAs, one has to ask which classes of events (geological, biological, atmospheric, oceanic, or human) have been significant in the history of earthworms in the area. For the LAs it is clear that there has been considerable movement of earthworms due to human activity. Much of this probably took place since 1492, but we cannot rule out the possibility of horticultural commerce and earthworm transport prior to 1492. The latter is important to our concerns because it would allow a much greater time period for the spread of exotic species within and among islands. Thus it would be possible to find exotic worms deep in the interior forests and mountains, in regions where no recent cultivation has taken place.

Perhaps one can dismiss atmospheric dispersal of earthworms, and one may well question the likelihood of dispersal via salt water, except by rafting. There are salt water-tolerant species of earthworms on sub-antarctic islands, but they are not believed to be ancestral to any secondarily fresh water or soil dwellers

(Lee 1959). Even so, the prevailing currents used to take ships from the LAs towards Venezuela rather than away from it. A prevailing ocean current sweeps northwestward across the NE face of South America and into the Caribbean Sea and the Gulf of Mexico (Richardson and Walsh 1986). Circulation within the Caribbean would be in the clockwise direction, bringing water from north to south on the west side of the LAs arc and feeding into the northwest-bound current. From this unsophisticated look at ocean currents, one could conclude that rafting would be more likely to take earthworms from the LAs to South America rather than vice-versa. Rafting however, would be possible from the Greater Antilles to the LAs. Nevertheless, prevailing easterly winds would not favor any rafting to any of the Antilles. Only large cyclonic storms offer the possibility of rafting in the desired directions, since such storms would disrupt the ordinary circulation patterns of air and water. On the other hand, proximity of the LAs to one another and to land masses with earthworms may have allowed some natural dispersal to take place.

If non-human dispersal of earthworms appears unlikely, could there have been any points in the geological evolution of the LAs at which earthworm invasion was probable? The well-documented portions of the geological history of the LAs suggest they would not have naturally-occurring earthworm faunas.

The Lesser Antilles are believed to have originated from volcanic action at the zone of subduction of the Atlantic Plate under the Caribbean Plate (Burke et al. 1984, Maury et al. 1990). Apparently there has been no contact with other land masses during the evolution of the archipelago. There are two lines of evidence supporting mechanisms besides oversea dispersal: the basement rock of La Desirade (near Guadeloupe) and probably of the adjacent islands is 140–150 million years old (Maury et al. 1990), and Roughgarden (1990) suggests that *Anolis* lizard data support a hypothesis of contact with terranes derived from the Greater Antilles. If it is true that a fragment of land was left in the wake of the separation of Africa from North and South America, it could have retained some earthworms. In this case one would predict closer affinities of the worms to their African relatives than to Neotropical relatives. This remains to be evaluated.

On the other hand, if a small fragment of land from an American source is responsible for the presence of these earthworms on the three islands, one would expect greater affinity to other Neotropical species. Again, this awaits detailed study of the material from Dominica, Guadeloupe, and Martinique. However, *Andioscolex* is clearly neotropical, belonging to the South American family Glossoscolecidae. *Andioscolex peregrinus* was described from specimens taken from plants shipped to Hamburg from the West Indies (Michaelsen 1897), and has since been found in Bolivia (Cognetti 1902). The new species found on Dominica is not *A. peregrinus*, and the exact origin of the plants is not known. A species from Martinique is also different from any described members of the genus and from the Dominican species. Since the volcanic origin of the LAs took place in mid-ocean, conventional wisdom states that the biota of these islands arrived by air- or water-borne dispersal. Other mid-oceanic volcanic archipelagoes, such as the Canary Islands (Talavera 1990) and the Hawaiian Islands (Nakamura 1990, Loope et al. 1988) lack indigenous earthworm species, although they have acquired abundant populations of exotics.

3.   Greater Antilles

The Greater Antilles (GAs), including the Virgin Islands, are home to diverse earthworm faunas related to taxa found in Mexico, Central America, South America, and Africa. Only Puerto Rico has been surveyed fairly well. Few surveys have been done in Cuba. In contrast to the LAs, the majority of geological models for the evolution of the GAs offer some indications for past direct contact with continents. In this section we will first quickly mention the general models of the origin of the Greater Antilles, and then review the earthworm taxa present, integrating relevant points from the geology.

Northern Hispaniola, Puerto Rico, and the Virgin Islands are thought to have originated as a volcanic island arc in the eastern Pacific (Burke et al. 1984, Duncan and Hargraves 1984). The arc rode the Caribbean plate through a gap between "nuclear Central America" (roughly that north of modern Nicaragua) and Colombia, arriving in its present location about 20 million years ago. Cuba is partly of continental origin but moved in concert with the other members of the northern GAs (Lewis and Draper 1990). Jamaica, southwestern Hispaniola, and perhaps eastern Cuba are hypothesized to have originated as a separate arc that arrived later (Burke et al. 1984, Lewis and Draper 1990). Geologists are not in complete agreement on the geological history of the GAs (see Pindell and Barrett 1990), or on the details of the histories of certain islands (cf. Pregill 1981). Our purpose is to give some preliminary indications of where the earthworm distributions and the geological evidence agree and disagree.

The first anomaly in the fauna is the lack (so far) in Hispaniola of any species related to the *Trigaster* group of genera (*Borgesia, Neotrigaster, Zapatadrilus, Trigaster*) (see Appendix 1). According to the geological models mentioned above, Hispaniola would be the corridor from Cuba to Puerto Rico and the Virgin Islands. Members of the *Trigaster* group (James 1991) inhabit all but Hispaniola. Their absence from Hispaniola is a challenge. Two possible explanations that preserve the model are (1) *Trigaster* were present but went extinct following the arrival of *Dichogaster-Eutrigaster* species on the southwest section of Hispaniola, or (2) *Trigaster* are present but have not been found or were present in recent times but have been driven to extinction by habitat destruction. How can one know that *Dichogaster* spp. are late arrivals? One bit of evidence is their absence from Puerto Rico, which did not receive any terranes at the time Jamaica, southwestern Hispaniola, and eastern Cuba are thought to have moved into the Caribbean Sea. The presence of *Dichogaster* and its relatives in Hispaniola and Jamaica, and the recent studies by Rodríguez (personal communication) who found *Eutrigaster* in eastern Cuba (Guantanamo), support the first explanation (displacement hypothesis). A more intensive sampling in the islands will bring the final answer.

Another anomaly is the lack in the Virgin Islands of glossoscolecids similar to the Puerto Rican genus *Estherella*. The closest relatives of *Estherella* live in Colombia (Figure 1a), supporting an ancient connection to South America. On the other hand the Virgin Islands seem to be entirely occupied by one or more species of *Trigaster*, a genus also present in Puerto Rico. The Virgin Islands, except for St. Croix, are part of the Puerto Rico Bank geological unit (Lewis and Draper 1990). St. Croix is unique among the Virgin Islands in that it lacks an endemic earthworm fauna, though many exotic species are present (Gates 1982, S.W. James unpublished data). St. Croix is also geologically unique among the Virgin Islands, its rocks having originated primarily as oceanic sediments (Lewis and Draper 1990). For the present it appears that St. Croix's origins did not admit a native earthworm fauna, and there has been no over-water dispersal of earthworms aside from that caused by human activity in recent times.

Finally the native earthworms of Jamaica (composed of *Dichogaster* and *Eutrigaster*) are related to the fauna of Cuba, Mexico, and Central America (see Appendix 1). Geological models suggest that Jamaica was once connected to Central America, and probably acquired its earthworm fauna at that time. Much of Jamaica was inundated during the Eocene (Lewis and Draper 1990), but unless there was over-water dispersal after the inundation(s), the entire island was never simultaneously inundated.

Considering the earthworm taxa present on the various GAs, it is clear that there must have been some communication between these islands and various continental areas. Puerto Rico shows evidence of past proximity to northern South America, Jamaica with Costa Rica, and Cuba, Puerto Rico, the Virgin

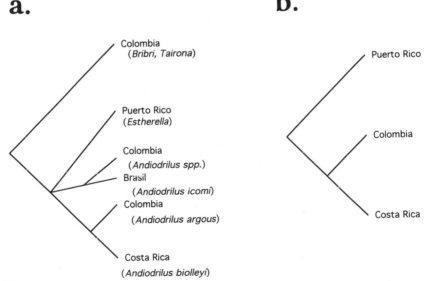

Figure 1. a. Area cladogram for *Estherella* and related genera. b. Predicted area cladogram for *Estherella* and *Andiodrilus*.

Islands and Hispaniola with Central America and/or Mexico. All of the relationships are consistent with a model of GAs evolution that considers the main arc (Cuba-Virgin Islands) as a single mass that connected to South America at the Virgin Islands end and to some part of nuclear Central America at the Cuba end, the two connections not necessarily occurring simultaneously.

Following are the three proposed modifications of conventional interpretations of Caribbean geology and biogeography:

1. South American earthworm stock entered Puerto Rico during the Cretaceous, as opposed to via the Lesser Antilles, as have other taxa. The distribution of the Glossoscolecidae provides evidence for colonization of Puerto Rico from South America. The Puerto Rican genus *Estherella* closely resembles Colombian *Andiodrilus*. The former is hypothesized to have invaded the Proto-Antilles from the south (modern east), since new Mexican Glossoscolecidae are very different and thus exclude the possibility of an invasion from the north. Contiguity of at least the Puerto Rico Bank is required to enable *Estherella*'s ancestors to arrive over what is now the Virgin Islands.

   The taxon/area cladogram in Figure 1a was derived from data on the Puerto Rican *Estherella* (Borges and Moreno 1989, James unpub.) and three related Colombian genera, *Andiodrilus* (Righi 1971, 1984, Cognetti 1904a, Zicsi 1988), *Tairona* and *Bribri* (Righi 1984), using the METRO parsimony routine in the PHYLIP package Ver. 2.8 (Felsenstein 1985). Figure 1b is the predicted area cladogram based on a hypothetical colonization of Proto-Puerto Rico from Colombia, followed by colonization of Panama and Costa Rica from the same source. With inadequate data on most taxa and only one taxon from Costa Rica in the data set, this model cannot be definitively evaluated, but the analysis does support close affinity of *Estherella* and *Andiodrilus*.

2. The Proto-Antilles formed a continuous isthmus or island at some point, in contrast to geological models of Coney (1982) and Burke et al. (1984), once connected to northern and southern continents (though not necessarily simultaneously). Mexican ancient *Zapatadrilus* invading the Proto-Antilles from the north must have had continuity in order to reach Puerto Rico and the Virgin Islands. The *Trigaster* data from James (1991) were used to create the cladagram in Figure 2 (made with the same technique as in Figure 1), which demonstrates the affinities of Puerto Rico and the Virgin Islands to North Mexico. This second modification of Rosen's model (1975, 1985) is also supported by data on plants and herpetofauna (Savage 1982). In this context we must try to clarify why *Zapatadrilus* and *Zapotecia* are absent in northern Central America and southern Mexico, but present in Cuba and Hispaniola, and why the genera *Lavellodrilus* (Acanthodrilini) and *Ramiellona* (Dichogastrini) are absent in the GAs but present in Central America and southern Mexico. This could be satisfactorily explained assuming the drifting of a large portion of land during the Cretaceous. This will correspond to the separation of the Chortis block from North America, during a period of nearly 80 million years (Donnelly 1988). According to this author the Chortis

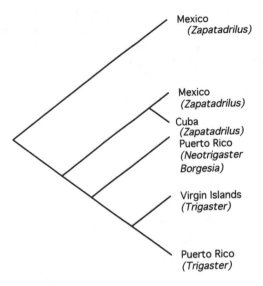

Figure 2. Area cladogram for *Trigaster* and related genera.

block includes southern Guatemala, Honduras, and most of Nicaragua. However, the biogeographical data on earthworms suggest that this block should be greater, probably including all of Guatemala, El Salvador, and the southern Mexican states of Chiapas, Oaxaca, and Tabasco.

3. According to the geological record, Jamaica was almost completely submerged at one or more times since its departure from Central America (Buskirk 1985). Yet Jamaica has numerous endemic earthworms, and *Pelmatodrilus*, an enchytraeid oligochaete ectocommensal on native Jamaican earthworms (Coates 1990, James pers. observ.). Without a fairly esoteric dispersal scenario, it is difficult to imagine such a diverse fauna dispersing to an island that was inundated even once since its departure from Central America. Provisionally we hypothesize that Jamaica has never been completely inundated since it severed its connections with a source of earthworms, among other biota.

C. Agriculture, Disturbance, and the Fate of Native Species

Eisen (1900) was the first to call attention to the disappearance of native species in disturbed systems of the north neotropical region. In the Caribbean, Michaelsen (1935) also noticed that native earthworms were vanishing from regions influenced by human activity, and were being replaced by exotic species from Asia, Africa, and South America. By now it is widely known that exotic species dominate in disturbed ecosystems, with native species predominating in natural, undisturbed environments (New Zealand, Lee 1961; North America, Smith 1928, Kalisz and Dotson 1989; South Africa, Ljungstrom 1972). It is also recognized that only a few groups of exotic species ("peregrines" of Lee 1987), predominate in disturbed lands. For biodiversity conservation practices, the main point is to evaluate the chance of native species to survive once their natural environment is disturbed. Here we present some examples from Mexico and the Caribbean Islands that suggest that native species have some opportunity to survive, even in disturbed ecosystems.

In Mexico the relationships between native and exotic species have been studied at the local and regional scale by Fragoso (1992, 1993), Fragoso et al. (1993) and Rojas and Fragoso (1994). At the regional scale the results indicate that the number of native species is significantly higher in natural sites (3.6 species) than in disturbed ones (0.96 species) (Fragoso et al. 1993) and that the number of exotic species is similar in both systems (ca. one species). In addition, of the native species recorded in southeastern Mexico (69 species), 62% were exclusively found in natural ecosystems (Fragoso 1993). Considering the deforestation rates of Dirzo and García (1992) and that nearly 53% of southern Mexico's surface area has been disturbed, it is possible to anticipate a large extinction of presently surviving native earthworms, unless their habitat is preserved or (as it will be seen later) the use of conservative agricultural practices is generalized.

In the same study, Fragoso (1993) found some native species capable of surviving in disturbed ecosystems, which in general corresponded to widely distributed species (*Balanteodrilus pearsei, Lavellodrilus parvus, Diplotrema murchiei*; Figure 3). These were eurytopic species (tolerant of different climatic and edaphic variables) with a marked preference for pastures. They can thus be considered as local peregrine species.

Fragoso (1993) discussed the factors that determine the survival of native species in southeastern Mexico, and arrived at the following conclusions:

1. Both in natural and disturbed ecosystems, earthworm communities are composed, to a minor or major extent, of a mixture of native and exotic species.
2. In general, communities in natural ecosystems have lower abundancesand biomasses than those in disturbed sites, with pastures presenting the greater values.
3. Epigeic and polyhumic species are more affected by perturbation than are endogeics.
4. In disturbed ecosystems the survival of native species depends on the number of years that a site has remained disturbed, the intensity of destructive agricultural practices (tillage, pesticides, etc.) and the ecological category of the species.
5. When natural vegetation is eliminated, the earthworm communities change; in some situations the original species are capable of adapting to the new conditions but often they fail and become locally extinct. Alien species (exotic and native from other sites) may then invade the community, their number

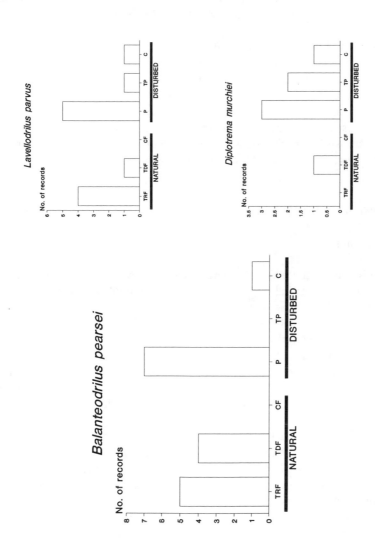

Figure 3. Distribution of widespread native earthworms in natural and disturbed ecosystems of tropical eastern Mexico. TRF=tropical rain forests, TDF=tropical subdeciduous forests, CF=cloud forests, P=pastures, TP=tree plantations, C=crop cultures (taken from Fragoso 1993).

and abundance being determined by the empty niches left by the disappearance of the native species (also see Kalisz and Wood this volume).

S. Borges (unpublished data) found in Puerto Rico native species of the
genera *Estherella*, *Borgesia*, *Neotrigaster*, and *Trigaster* exclusively inhabiting
natural sites; exotics, on the other hand, were limited to disturbed ecosystems.
Only three native species of the genus *Pontoscolex* were frequently found in
disturbed sites. Thus, the wide physiological plasticity observed in *P.
corethrurus* (Lavelle et al. 1987) could be a genetic feature.

In the Virgin Islands (U.S. and British), the large and often numerous
Trigaster species were found in a wide climatic range of natural and disturbed
habitats, including degraded pasture, thorn scrub, secondary growth forests, and
old growth moist forests (James unpub.). These worms could be a useful
addition to sustainable agriculture research in the region. In the Dominican
Republic native species of *Dichogaster* and *Temanonegia* were found in a similar
variety of habitats, though in a less extreme climatic range (James unpub.).

III.   Research Imperatives

A.   Taxonomy

Undoubtedly, one of the major challenges that faces earthworm research in the
region is to complete and clarify the taxonomy. This implies a need for a great
sampling effort in the near future. Moreover, taxonomists should pay special
attention to solving taxonomic puzzles. The following is a list of the major
problems that we recognize for the region.

1. **Intensive sampling**. Data formerly presented revealed that in several
   countries taxonomic research on earthworms is very scarce. Obviously this
   must be prioritized. In Mexico, for example, only the east and southeast
   regions have been moderately well sampled, and even in these regions there
   are several unsampled sites. If we look at the sampling effort in this country
   thus far, it is soon realized that the major part of the country is still unknown.
   Sampling must be focused on the west coast and central and western mountain
   systems, particularly at river edges, remnants of natural vegetation and
   pastures. The magnitude of the taxonomic ignorance is appreciated when the
   earthworm fauna of this country is compared with India, a tropical country
   with a similar area: India has more than 400 described species, whereas
   Mexico only presents 115 species.

   Central America is practically unexplored, and in two countries (Belize and
   Nicaragua) there are no records of native earthworms. As in Mexico,
   sampling must also be conducted in natural ecosystems. Special attention must
   be paid to comparisons between high and low altitude lands of Guatemala,
   Honduras, and Nicaragua.

In the Caribbean the work recently made by James and Borges and Moreno has improved our knowledge of these islands. There are, however, large gaps to be filled, especially in the Greater Antilles (Cuba and Hispaniola).

2. **Phylogenetic affinities**. Only after intensive sampling has been accomplished will it be possible to analyze phylogenetic relationships, as has been done with some taxa found in this region (James 1991, Borges 1992). The following are several examples that illustrate specific problems requiring clarification.

In the Caribbean the relationships between Puerto Rican *Estherella* and Colombian *Andiodrilus* must be clarified. One may question if the differences between the two genera are really of generic significance.

In Mexico and Central America it is particularly necessary to clarify the relationships between existing and newly described taxa of Acanthodrilini and Dichogastrini.

Special attention is deserved by the complex *Dichogaster-Eutrigaster*. *Dichogaster* (s.l.) is a large genus (>200 species) with many artificially dispersed species (Gates 1972). Representatives are found naturally in Africa, Southeast Asia, the South Pacific, South America, Central America, Mexico, Hispaniola, Jamaica, Martinique, Dominica, and Guadeloupe. Endemics, however, are restricted to Mexico, Central America, the Caribbean islands, South America and Africa. Recent changes proposed by Sims (1987) and Csuzdi and Zicsi (1991) (resurrection of *Eutrigaster* to include all the American species of *Dichogaster* with posterior first dorsal pore and proventriculus) constitute a laudable effort, but beg the question what to do with the scores of species from Africa, Asia, and the south Pacific. *Dichogaster* is a large group, and it seems that much revision has been left unattempted in the interest of maintaining stability of nomenclature. It also seems that those interested in earthworms of the New World may have to tackle an area larger than the neotropics in order to achieve a sensible classification of the western hemisphere taxa. We recommend that this genus, its close relatives, and several species from Africa (Omodeo 1958) be subjected to close scrutiny with major revision as the goal.

The reorganization of classical *Trigaster* proposed by James (1991) (*Neotrigaster* and *Borgesia* exclusively Puerto Rican, *Trigaster* in the Virgin Islands plus Puerto Rico, and *Zapatadrilus* in Mexico and Cuba) is provisional, and will need more attention when more Cuban material is available. Now it is important to clarify the status of *Trigaster* (material collected by James from numerous Virgin Islands will contribute towards a revision of this genus) and the relationships between Mexican and Cuban *Zapatadrilus*.

We believe the neotropical Megascolecidae (Acanthodrilini and Dichogastrini) to be very diverse, and the species presently known to be a small fraction of the total. It is premature to state that the group is ripe for revision, but it may be better to lay a structure now, and allow the taxonomic work to fill it up, than to try and "rearrange a haystack" in ten years. Therefore one research priority should be to revise the Neotropical representatives and any closely related genera from other areas.

The Ocnerodrilinae are an obscure group, most of which are quite small. The recent finding in Mexico of a new malabarin genus (Fragoso 1993) indicates that more species must exist, either in continental or insular ecosystems.

The presence of native Glossoscolecidae in southern Mexico, enables us to anticipate the occurrence of more species in other Central American countries.

3. **Facilitation of earthworm taxonomy**. In addition to clarifying the classification system and eliminating gaps in knowledge, there are some "applied" science projects that should be undertaken. Taxonomists should look into making the study of earthworms easier for everyone. Here are some suggestions:

- Publish an introductory volume or paper with anatomical maps of the families of New World earthworms and exotics found here.
- The same volume/paper should include examples of minimum standards for species description, and taxonomists should agree to abide by those standards.
- Keys to the known species are very much needed for nearly every group. The last comprehensive key to the Oligochaeta was published in 1900. It is beyond our interests to write keys to the entire Oligochaeta, but we should at least get to work on the groups we know. The keys should be accompanied by brief definitions, as in Michaelsen (1900), to prevent misidentifications. For ease of updating, it may be useful to also render the key in a computerized form, such as the Hypercard program generated by Colosimi et al. (1991). The "key" would take the form of a database with a matching built-in algorithm. There may be many ways to accomplish this. The overall advantage is the ease of distribution and revision of the database, such as over computer networks.

B.   Biodiversity and Conservation

When thinking about biodiversity conservation it is important to keep in mind that more species may be found in the future, or that there may be indigenous species on an island or region thought to lack them. Absence of data is always less convincing than the presence of it, after all, one may have missed what is there, but it is hard to argue against what has actually been collected. Thus absence of data in a region or island may be a result of habitat destruction and/or species extinction. Extinction could be local, in the more accessible areas, or global, over the entire species range. Logically, we can never know what once existed, and therefore we can never totally reject the hypothesis that earthworms did reach a particular island or region by natural means.

The rapid pace of habitat destruction in the Neotropics is widely recognized as a threat to biotic diversity. In the interests of learning about and potentially preserving biotic diversity, Caribbean, Mexican, and Central American earthworms need to be studied soon. In this regard sampling should be mainly directed to the areas most likely to harbor native earthworm species. This will require extensive cooperation among earthworm specialists and those most familiar with the remaining natural habitats (local scientists and foresters).

To fully understand the extinction of natives, we must give special attention to the spread of peregrine species, and to their negative effect on indigenous earthworm species populations (see Kalisz and Wood this volume). Evidence shown in this paper and fully developed by Fragoso (1993) suggest that in Mexico only a few native species are capable of surviving and coexisting with exotic worms in disturbed ecosystems.

In natural environments, on the other hand, is it probable that exotic species displace native ones? What kind of interactions exist between native and exotics? If competition occurs, does it result in exclusion? Is the outcome of competition different in large and small reserves of natural vegetation? If so, why? Will peregrine species eventually cover the entire Neotropics? Fragoso (1993) has found little evidence that this is occurring, even in small reserves. In the Mexican east tropics small and large reserves can maintain a great deal of their original native earthworm fauna. Nevertheless, more research is needed to clarify these points, especially when results in temperate zone forests suggest that small isolated reserves will not retain native earthworm populations in the face of pressure from exotics (Kalisz and Dotson 1989, Dotson and Kalisz 1989)

C.   Ecological Studies

Results presented in this paper point out that data on earthworm ecology in the NNR are very scarce. Thus ecological studies must be encouraged, considering that: (1) in the region there are still several types of once abundant, natural vegetation, and (2) earthworms dominate, in terms of fresh biomass, the soil biota of natural and disturbed soils of the humid tropics (Lavelle et al. 1994, in press).

Particularly relevant to tropical forest dynamics will be studies on the contributions of earthworms to litter decomposition, nutrient recycling and soil properties.

A primary difficulty in determining what effects earthworms have is connected to the means necessary to remove them for experimental purposes. Biocide applications do not affect just one organism, and it may take years for the residual effects of the exterminated worms to disappear. Mechanical removal is possible but the cost is very expensive and contamination by alien species could always occur. In the LAs there is presently a variety of sites at different elevations with and without earthworms, and all under slightly disturbed or undisturbed vegetation. With care, it may be possible to conduct comparative studies that will yield interesting results. Given the volcanic origins of the soils on many islands and the relatively similar vegetation, it may be possible to pair up sites that differ primarily in the presence or absence of native earthworms. Moreover within an island devoid of native species, it is possible to find sites with and without introduced species of earthworms. Thus it may be possible to determine the effects of earthworm introductions on forest ecology, the structure of soil, and the composition of the soil biota.

Acknowledgments

The authors wish to thank Dr. Paul Hendrix for the invitation and resources to participate in the workshop. We also express our sincere gratitude to all the staff of the Institute of Ecology, University of Georgia, at Athens. C. Fragoso also wishes to acknowledge the financial support of the EEC (Projects TS2 *0292-F and ERBTS3*CT920128). S.W. James research was supported by the U.S. National Science Foundation.

References

Alfaro, M. 1992. *Oligoquetos terrestres de la Laguna Cartagena y algunos aspectos de su ecología*. M.S. thesis, Univ. Puerto Rico, Mayagüez. 103 pp.

Atkin, L. and J. Proctor. 1988. Invertebrates in the litter and soil on Volcán Barba, Costa Rica. *J. Trop. Ecol.* 71:503–527

Borges, S. 1988. *Los oligoquetos terrestres de Puerto Rico*. Ph.D. thesis, Univ. Complutense, Madrid, Spain. 376 pp.

Borges, S. 1992. A tentative phylogenetic analysis of the genus *Pontoscolex* (Oligochaeta: Glossoscolecidae). *Soil Biol. Biochem.* 24:1207–1211.

Borges, S. (in press a). A new species of *Onychocha*eta Beddard, 1891 (Oligochaeta: Glossoscolecidae) from Puerto Rico. *Carib. J. Sci.*

Borges, S. (in press b). A study of the earthworm diversity and ecology in the Bisley Watersheds and El Verde, Luquillo Experimental Forest. Abstract. IITF Newsletter

Borges, S. and A.G. Moreno. 1989. Nuevas especies del género *Estherella* Gates, 1970 (Oligochaeta: Glossoscolecidae) para Puerto Rico. *Boll. Mus. Reg. Sci. nat. Torino* 7(2):383–399.

Borges, S. and A.G. Moreno. 1990a. Contribución al conocimiento de los oligoquetos terrestres de Puerto Rico: las "pheretimas". *Carib. J. Sci.* 26:141–151.

Borges, S. and A.G. Moreno. 1990b. Nuevas especies y un nuevo subgénero del género *Pontoscolex Schmarda,* 1861 (Oligochaeta: Glossoscolecidae) para Puerto Rico. *Boll. Mus. Reg. Sci. nat. Torino* 8(1):143–157.

Borges, S. and A.G. Moreno. 1991. Nuevas especies del género *Trigaster* Benham, 1886 (Oligochaeta: Octochaetidae) para Puerto Rico. *Boll. Mus. Reg. Sci. nat. Torino* 9:39–54.

Borges, S. and A.G. Moreno. 1992. Redescripción de *Trigaster rufa* Gates, 1962 (Oligochaeta: Octochaetidae). *Carib. J. Sci.* 28:47–50.

Borges, S. and A.G. Moreno. 1994. Dos citas nuevas de oligoquetos para Puerto Rico, y nuevas localidades para otras tres especies. *Carib. J. Sci.* 30(1–2):150–151.

Burke, K., C. Cooper, J.F. Dewey, P. Mann, and J.L. Pindell. 1984. Caribbean tectonics and relative plate motions. *Geol. Soc. Am. Mem.* 162:31–63.

Buskirk, R.E. 1985. Zoogeographic patterns and tectonic history of Jamaica and the northern Caribbean. *J. Biogeography* 12:445–461.

Campa, L., M. Ojeda, V. Vale, and J.R. Cuevas. 1990. Desarrollo de una lombricultura tropical para el procesamiento de los residuales sólidos orgánicos. Abstract. XI Congreso Latinoamericano de la Ciencia del Suelo, La Habana, Cuba.

Coates, K. 1990. Redescriptions of *Aspidodrilus* and *Pelmatodrilus*, enchytraeids (Annelida: Oligochaeta) ectocommensal on earthworms. *Can J. Zool.* 68:498–505.

Cognetti, L. 1902. Viaggio del Dr. Borelli nel Chaco boliviano e nella Republica Argentina. XVII. Terricole boliviane ed Argentini. *Boll. Mus. Torino* 17(420):1–11.

Cognetti, L. 1904a. Oligocheti di Costa Rica. *Boll. dei Mus. di Zool. Anat. Comp. della R. Univ. Torino* 19(462):1–10.

Cognetti, L. 1904b. Nuovi Oligocheti di Costa Rica. *Boll. dei Mus. di Zool. Anat. Comp. della R. Univ. Torino* 19(478):1–4.

Cognetti, L. 1906. Gli Oligocheti della Regione Neotropicale, II. *Mem. R. Accad. Sc. Torino* 56:147–262.

Cognetti, L. 1907. Nuevo contributo alla conoscenza della drilofauna neotropicale. *Atti. R. Accad. Sc. Torino* 42:789–800.

Cognetti, L. 1908. Lombrichi di Costa Rica e del Venezuela. *Atti. R. Accad. Sc. Torino* 43:505–518.

Colosimi, A., R. Rota, and P. Omodeo. 1991. A Hypercard program for the identification of biological specimens. *Cabios* 7:63–69.

Coney, P.J. 1982. Plate tectonic constraints on the biogeography of Middle America and the Caribbean region. *Ann. Mo. Bot. Gard.* 69:432–443.

Csuzdi, C. and A. Zicsi. 1991. Über die Verbreitung beuer und bekannter *Dichogaster* und *Eutrigaster* arten aus Mittel- und Südamerika (Oligochaeta: Octochaetidae). Regenwürmer aus Südamerika 15. *Acta Zool. Hung.* 37(3–4):177–192.

Darwin, C. 1881. *The formation of vegetable mould, through the action of worms, with observations on their habits.* Faber and Faber, London.

Dirzo, R. and M.C. García. 1992. Rates of deforestation in Los Tuxtlas, a neotropical area in southeast Mexico. *Conservation Biol.* 6:84–90.

Donnelly, T.W. 1985. Mesozoic and Cenozoic plate evolution of the Caribbean regions. p. 89–121. In: Stehli, F.G. and D. Webb (eds.) *The great American biotic interchange.* Plenum Press, New York.

Donnelly, T.W. 1988. Geological constraints on caribbean biogeography. p. 15–37. In: Liebherr, J.K. (ed.) *Zoogeography of Caribbean insects.* Cornell Univ. Press. Ithaca, NY.

Dotson, D.B. and P.J. Kalisz. 1989. Characteristics and ecological relationships of earthworm assemblages in undisturbed forest soils in the southern Appalachians of Kentucky, USA. *Pedobiologia* 33:211–230.

Duncan, R.A. and R.B Hargraves. 1984. Plate tectonic evolution of the Caribbean region in the mantle reference frame. *Geol. Soc. Am. Mem.* 162:81–93.

Eisen, G. 1895. Pacific Coast Oligochaeta I. *Mem. Calif. Acad. Sci.* 2:63–122.

Eisen, G. 1896. Pacific Coast Oligochaeta II. *Mem. Calif. Acad. Sci.* 2:123–198.

Eisen, G. 1900. Researches in American Oligochaeta, with especial reference to those of the Pacific coast and adjacent islands. *Proc. Cal. Acad. Sci. 3rd Ser. Zoology* 2:85–276.

Felsenstein, J. 1985. Confidence limits on phylogenies: an approach using the bootstrap. *Evolution* 29:783-791.

Ferrusquia, V.I. 1993. Geology of Mexico: a synopsis. p. 3-107. In: Ramamoorthy, T.P., R. Bye, A. Lot,. and J. Fa (eds.) *Biological diversity of Mexico: origins and distribution.* Oxford University Press.

Flores, O. and P. Gerez. 1988. *Conservación en México: Síntesis Sobre Vertebrados Terrestres, Vegetación y Uso del Suelo.* INIREB-Cons. International. México.

Fragoso, C. 1985. *Ecología general de las lombrices terrestres (Oligochaeta: Annelida) de la región Boca del Chajul, Selva Lacandona, Estado de Chiapas.* Thesis. Univ. Nacional Autónoma de México, México. 133 pp.

Fragoso, C. 1988. Sistemática y ecología de un género nuevo de lombriz de tierra (Acanthodrilini: Oligochaeta) de la Selva Lacandona, Chiapas, México. *Acta Zool. Mex.* 25:1-39.

Fragoso, C. 1989. Las lombrices de tierra de la Reserva El Cielo. Aspectos ecológicos y sistemáticos. *Biotam* 1(1): 38-44.

Fragoso, C. 1990. Las lombrices de tierra (Oligochaeta, Annelida) de la península de Yucatán. p. 151-154. In: Robinson, J.G. and D. Navarro (eds.) *Diversidad biológica en Sian Ka'an, Quintana Roo, México.* Centro de Investigaciones de Quintana Roo. Program for studies in tropical conservation. Univ. Florida, Gainesville.

Fragoso, C. 1991. Two new species of the earthworm genus *Lavellodrilus* (Oligochaeta, Acanthodrilini) from tropical Mexican rain forests. *Studies Neotrop. Fauna Environment* 26(2):83-91.

Fragoso C. 1992. Las lombrices terrestres de la Selva Lacandona, Ecología y Potencial Práctico. p. 101-118. In: Vásquez-Sánchez, M.A. and M.A. Ramos (eds.) *Reserva de la Bíosfera Montes Azules, Selva Lacandona: Investigación para su uso.* Publ. Esp. Ecósfera 1.

Fragoso C. 1993. *Les peuplements de vers de terre dans l'est et sud'est du Mexique.* Ph.D. thesis. Univ. Paris 6. Paris, France. 225 pp.

Fragoso, C. and P. Lavelle. 1987. The earthworm community of a Mexican tropical rain forest (Chajul, Chiapas). p. 281-295. In: Bonvicini, A. M. and P. Omodeo (eds.) *On earthworms.* Selected Symposia and Monographs U.Z.I., 2, Mucchi, Modena, Italy.

Fragoso, C. and P. Lavelle. 1992. Earthworm communities of tropical rain forests. *Soil Biol. Biochem.* 24:1397-1408.

Fragoso, C., I. Barois, C. González, C. Arteaga, and J.C. Patrón. 1993. Relationship between earthworms and soil organic matter levels in natural and managed ecosystems in the Mexican tropics. p.231-239. In: Mulongoy, K. and R. Merckx (eds.) *Soil organic matter dynamics and sustainability of tropical agriculture.* Wiley-Sayce Co-Publication. Sussex, U.K.

Fragoso, C. and P. Rojas. 1994. Earthworms from southeastern Mexico. New acanthodriline genera and species (Megascolecidae, Oligochaeta). *Megadrilogica* 6(1):1-12.

Fraile, J.M. 1989. Poblaciones de lombrices de tierra (Oligochaeta: Annelida en una pastura de *Cynodon plectostachyus* (pasto estrella) asociada con árboles de *Erythrina poeppigiana* (puro), una pastura asociada con árboles de *Corida alliodora* (laurel), una pastura sin árboles y vegetación a libre crecimiento en el CATIE, Turrialba Costa Rica. M.Sci. thesis. Univ. de Costa Rica y CATIE.

Gates, G.E. 1942. Checklist and bibliography of North American earthworms. *Am. Midl. Nat.* 27:86-108.

Gates, G.E. 1954. Exotic earthworms of the United States. *Bull. Mus. Comp. Zool. Harvard* 3(6): 217–258.

Gates, G.E. 1957. Contribution to a revision of the earthworm family Ocnerodrilidae. The genus *Nematogenia. Bull. Mus. Comp. Zool.* 117:427–445.

Gates, G.E. 1962a. On some earthworms of Eisen's collections. *Proc. Calif. Acad. Sci.* 4(31):185–225.

Gates, G.E. 1962b. On a new species of the earthworm genus *Trigaster* Benham 1886. (Octochaetidae) *Breviora* 178:1–4.

Gates, G.E. 1967. On a new species of earthworm from a Mexican cave. *Internatl. J. Speleol.* 3:63–70.

Gates, G.E. 1970a. On new species in a new earthworm genus from Puerto Rico. *Breviora* 356:1–11.

Gates, G.E. 1970b. On a new species of earthworm from another Mexican cave. *Southwest. Nat.* 15(2):261–273.

Gates, G.E. 1971. On some earthworms from Mexican caves. *Asoc. Mex. Cave Stud. Bull.* 4:3–8.

Gates, G.E. 1972. Burmese earthworms. An introduction to the systematics and biology of megadrile oligochaetes with special reference to southeast Asia. *Trans. Am. Phil. Soc.* 62:1–326.

Gates, G.E. 1973. On more earthworms from Mexican caves. *Asoc. Mex. Cave Stud. Bull.* 5:21–24.

Gates, G.E. 1977a. On some earthworms from North American caves. In: Studies on the caves and cave fauna of the Yucatan Península. *Asoc. Mex. Cave Stud. Bull.* 6:1–4.

Gates, G.E. 1977b. La faune terrestre de l'Ile de Saint-Helene. 1. Oligochaeta. *Mus. R. Afr. Cen. Ann. Belgique.* Ser. In 8°. Sc. Zool. 220:469–491.

Gates, G.E. 1978. On a new species of octochaetid earthworms from Mexico. *Proc. Biol. Soc. Wash.* 91(2):439–443.

Gates, G.E. 1979. A new genus of larger ocnerodrilid earthworms in the American hemisphere. *Megadrilogica* 3:162–164.

Gates, G.E. 1982. Farewell to North American megadriles. *Megadrilogica* 4:12–77.

Graff, O. 1957. Regenwürmer aus El Salvador. *Senck. Biol.* 38:115–143.

Hubers, H. 1993. *Los oligoquetos terrestres de los suelos Nipe (Oxisols) de Puerto Rico.* M.S. thesis, Univ. Puerto Rico, Mayagüez.

Humphries, C.J. and L.R. Parenti. 1986. *Cladistic Biogeography.* Clarendon Press, Oxford. xii + 98 p.

James, S.W. 1990. *Diplotrema murchiei* and *D. papillata*, new earthworms (Oligochaeta: Megascolecidae) from Mexico. *Acta Zool. Mex. (n.s.)* 38:18–27.

James, S.W. 1991. New species of earthworms from Puerto Rico, with a redefinition of the genus *Trigaster. Trans. Am. Micros. Soc.* 110(4): 337–353.

James, S.W. 1993. New acanthodriline earthworms from Mexico (Oligochaeta: Megascolecidae). *Acta Zool. Mex. (n.s.)* 60(1):1–21.

Jamieson, B.G.M. 1971. A review of the Megascolecoid earthworm genera (Oligochaeta) of Australia. Part I. Reclassification and checklist of the Magascolecoid genera of the world. *Proc. R. Soc. Qld.* 82:75–86.

Kalisz, P.J. and D.B. Dotson. 1989. Land-use history and the occurrence of exotic earthworms in the mountains of eastern Kentucky. *Am. Midl. Nat.* 122:288–297.

Ladd, J.W., T.L. Holcome, G.K. Westbrook, and N. T. Edgar. 1990. Caribbean marine geology: active margins of the plate boundary. p. 261–290. In: Dengo, G. and J.E. Case (eds.) *The Caribbean region.* Geol. Soc. Am., Boulder, CO.

Laessle, A.M. 1961. A micro-limnological study of Jamaican bromeliads. *Ecology* 42:499–517.

Lavelle, P., M. Maury, and V. Serrano. 1981. Estudio cuantitativo de la fauna del suelo en la región de Laguna Verde, Veracruz. Epoca de lluvias. In: Reyes-Castillo, P. (ed.) *Estudios ecológicos en el trópico mexicano.* Inst. Ecol. Publ. 6:65–100.

Lavelle, P. and B. Kolhmann. 1984. Etude quantitative de la macrofaune du sol dans une forêt tropicale mexicaine (Bonampak, Chiapas). *Pedobiologia* 27:377–393.

Lavelle, P., I. Barois, I. Cruz, C. Fragoso, A. Hernandez, A. Pineda, and P. Rangel. 1987. Adaptive strategies of *Pontoscolex corethurus* (Glossoscolecidae, Oligochaeta) a peregrin geophagus earthworm of the humid tropics. *Biol. Fertil. Soils* 5:188–194.

Lavelle, P. C. Gilot, C. Fragoso, and B. Pashanasi. 1994. Soil fauna and sustainable land use in the humid tropics. p. 291–308. In: Greenland, D.J. and I. Szabolcs (eds.) *Soil residue and sustainable land use.* CAB International. Wallingford, U.K.

Lavelle, P., M. Dangerfield, C. Fragoso, V. Eschenbrenner, D. Lopez, B. Pashanasi, and L. Brussaard. (in press). The relationship between soil macrofauna and tropical soil fertility. In: Woomer, P.L. and M.J. Swift (eds.) *The management of tropical soil biology and fertility.* Wiley-Sayce Publication. Sussex, U.K.

Lee, K.E. 1959. The earthworm fauna of New Zealand. *N. Z. Dept. Sci. Ind. Res. Bull.* 130. Wellington, New Zealand.

Lee, K.E. 1961. Interactions between native and introduced earthworms. *Proc. N.Z. Ecol. Soc.* 8:60–62.

Lee, K.E. 1987. Peregrine species of earthworms. p. 315–328. In: Bonvicini Pagliai, A. and P. Omodeo (eds.) *On earthworms.* Selected Symposia and Monographs, 2. Collana U.Z.I. Mucchi Editore, Modena, Italy.

Lewis, J.F. and G. Draper. 1990. Geology and tectonic evolution of the northern Caribbean margin. p. 77–140. In: Dengo, G. and J.E. Case (eds.) *The Caribbean region.* Geol. Soc. Am., Boulder, CO.

Ljungstrom, P.O. 1972. Taxonomical and ecological notes on the earthworm genus *Udeina* and a requiem for the South African acanthodrilines. *Pedobiologia* 12:100–110.

Loope, L.L., O. Hamann, and C.P. Stone. 1988. Comparative conservation biology of oceanic archipelagoes. *BioScience* 38:272–282.

López-Ramos, E. 1981. *Geología de México.* T. III 2da. Edición UNAM. México.

Maldonado-Koerdell, M. 1964. Geohistory and paleography of Middle America. p. 3–32 In: Wauchope R. (ed.) *Handbook of Middle American Indians.* Vol. 1. West, R. (ed.). *Natural Environment and Early Cultures.* Univ. of Texas Press, Austin.

Maury, R.C., G.K. Westbrook, P.E. Baker, PH. Bouysse, and D. Westercamp. 1990. Geology of the Lesser Antilles. p. 141–166. In: Dengo, G. and J.E. Case (eds.) *The Caribbean region.* Geol. Soc. Am., Boulder, CO.

Michaelsen, W. 1897. Organisation einiger neuer oder wenig bekannter Regenwurmer von Westindien und Sudamerika. *Zool. Jahrb. Anat.* 10:359–388.

Michaelsen, W. 1900. Oligochaeta. In: *Das Tierreich.* Lief. 10. Verlag von R. Friedlander und Sohn, Berlin.

Michaelsen, W. 1902. Neve Oligochaeta und neve fundorte altbekannter. *Mitt. Mus. Hamburg.* 19:1–54.

Michaelsen, W. 1908. Die Oligochäten Westindiens. *Zool. Jahrb Syst. Suppl.* 11:13–32.

Michaelsen, W. 1911. Zur Kenntnis der Eodrilaceen und ihrer Verbreitungsver hältnisse. *Zool. Jahrb. Syst.* 30:527–572.

Michaelsen, W. 1912. Über einige zentralamerikanische Oligochäten. *Arch. Naturgesch.* (*Abteilung A*) 78(9):112–129.

Michaelsen, W. 1923. Oligochäten von den warmeren Gebieten Amerikas und des Atlantischen Ozeans. *Mitt. Mus. Hamburgo.* 41:71–83.

Michaelsen, W. 1935. Die opisthoporen Oligochaten Westindiens. *Mitt. Mus. Hamburg* 45:51–64.

Michaelsen, W. 1936. African and American Oligochaeta in the American Museum of Natural History. *Am. Mus. Nov.* 843:1–20

Moore, A.M. and L. Burns. 1970. Appendix C: Preliminary observations on the earthworm populations of the forest soils of El Verde. p. 238. In: Odum, H.T. and R.F. Pigeon (eds.) *A tropical rain forest.* NTIS, Springfield, VA.

Nakamura, M. 1990. How to identify Hawaiian earthworms. *Chuo University Research Notes.* No. 11:101–110.

Omodeo, P. 1958. I. Oligochètes. In: La rèserve naturelle intègrale du Mont. Nimba. *Mèm. Inst. Fr. Afrique Noire* 4(53):9–109.

Pickford, E. 1938. Earthworms in Yucatan Caves. *Publ. Carnegie Inst. Wash.* 491:71–100.

Picado, C. 1913. Les broméliacées épiphytes considérées comme milieu biologique. *Bull. Sci. de la France et de la Belgique* 47:215–360.

Pielou, E.C. 1979. *Biogeography.* John Wiley & Sons, New York.

Pindell, J.L. and S.F. Barrett. 1990. Geological evolution of the Caribbean region; A plate-tectonic perspective. p. 405–458. In: Dengo, G. and J.E. Case (eds.) *The Caribbean region.* Geol. Soc. Am., Boulder, CO.

Pregill, G.K. 1981. An appraisal of the vicariance hypothesis of Caribbean biogeography and its application to West Indian terrestrial vertebrates. *Syst. Zool.* 30:147–155.

Reynolds, J.W. and G. Righi. 1994. On some earthworms from Belize, C.A. with the description of a new species (Oligochaeta: Acanthodrilidae, Glossoscolecidae and Octochaetidae). *Megadrilogica* 5(9):97–106.

Richardson, P. L. and D. Walsh. 1986. Mapping climatological seasonal variations of surface currents in the tropical Atlantic using ship drifts. *J. Geophysical Res.* 91:10537–10550.

Righi, G. 1971. Sobre a Familia Glossoscolecidae (Oligochaeta) no Brasil. *Arch. Zool. Est. S. Paulo* 20:1–96.

Righi, G. 1972a. Bionomic considerations on the Glossoscolecidae. *Pedobiologia* 12:254–260.

Righi, G. 1972b. On some earthworms from Central America (Oligochaeta). *Stud. Neotr. Fauna* 7:207–228.

Righi, G. 1984. On some earthworms (Oligochaeta, Glossoscolecidae) from the Sierra Nevada de Santa Marta (Colombia). *Stud. Trop. Andean Ecosystems* 2:455–468.

Righi, G. and J. Fraile. 1987. Alguns Oligochaeta de Costa Rica. *Rev. Brasil Biol.* 47 (4):535–548.

Rodríguez, C. 1991. Desarrollo del sistema reproductor y ovoposición de *Eudrilus eugeniae* (Oligochaeta: Eudrilidae) a 24 °C y 30 °C. *Revista Biol.* 5(2–3):159–167.

Rodríguez, C. 1993. Listado preliminar de las lombrices de tierra (Annelida: Oligochaeta) de Cuba. *Poeyana* 443:1–9.

Rodríguez, C., S. Brito, and A. Sierra. 1988. Estudio de los estados inmaduros de *Eudrilus eugeniae* (Oligochaeta: Eudrilidae) a dos temperaturas. *Revista Biol.* 2(2):45–54.

Rodríguez, C., M.E. Canetti, M. Reines, and A. Sierra. 1986. Ciclo de vida de *Eudrilus eugeniae* (Oligochaeta: Eudrilidae) a 30 °C. *Poeyana* 326:1–13.

Rodríguez, C. and A. González. 1986. Morfología y patrones electroforéticos de proteinas totales y esterasas de cuatro oligoquetos (Annelida: Oligochaeta) de Cuba. *Carib. J. Sci.* 22(1–2):71–83.

Rodríguez, C., G. Noriega, M. Reines, and A. Sierra. 1987. Estudio morfológico e identificación de un megadrilo (Oligochaeta) presente en Cuba. *Revista Biol.* 1(2):63–76.

Rodríguez, C. and M. Reines. 1986. Morfología de *Polypheretima elongata* (Oligochaeta: Megascolecidae) de una población cubana. *Poeyana* 325:1–10.

Rodríguez, C. and M. Reines. 1989. Algunas consideraciones acerca de la distribución mundial de *Polypheretima elongata* (Oligochaeta: Megascolecidae). *Ciencias Biol.* 21–22:98–103.

Rojas, P. and C. Fragoso. 1994. Fauna de Suelos del Edo. de Veracruz. Hormigas y lombrices de tierra en ecosistemas naturales y perturbados. p. 59–74. In: Gonzales, C.A. and A. Gonzales. *Problemática ambiental en el estado de Veracruz*. Fascículo Recursos Faunisticos, Xalapa, Mexico.

Rosen, D.E. 1975. A vicariance model of Caribbean biogeography. *Syst. Zool.* 24:431–464.

Rosen, D.E. 1985. Geological hierarchies and biogeographical congruence in the Caribbean. *Ann. Mo. Bot. Gard.* 72:636–659.

Roughgarden, J. 1990. Origin of the eastern Caribbean: data from reptiles and amphibians. In: LaRue, D. and G. Draper (eds.) *Transactions of the 12th Caribbean Conference*, St. Croix, U.S. Virgin Islands. Miami Geological Survey.

Savage, J.M. 1982. The enigma of the Central American herpetofauna: dispersals or vicariance? *Ann. Mo. Bot. Gard.* 69:464–547.

Sims, R.W. 1987. New species and records of earthworms from Jamaica with notes on the genus *Eutrigaster* Cognetti 1904 (Octochaetidae: Oligochaeta). *J. Nat. Hist.* 21:429–441.

Smith, F. 1928. An account of changes in the earthworm fauna of Illinois and a description of one new species. *Ill. Nat. His. Surv. Bull.* 17:347–362.

Stehli, F.G. and S.D. Webb. 1985. A kaleidoscope of plates, faunal and floral dispersals, and sea level changes. p. 3–15. In: Stehli, F.G. and D. Webb (eds.) *The great American biotic interchange*. Plenum Press, New York.

Talavera, J.A. 1990. Claves de identificacion de las lombrices de tierra (Annelida: Oligochaeta) de Canarias. *Vieraea* 18:113–119.

Zicsi, A. 1988. Neue *Andiodrilus*-arten aus Kolumbien (Oligochaeta: Glossoscolecidae). Regenwürmer aus Sudamerika 5. *Rev. suisse Zool.* 95:715–722.

Zicsi, A. and C. Csuzdi. 1991. Der erste Wiederfund von *Zapotecia amecamecae* Eisen, 1900 aus Mexico (Oligochaeta: Acanthodrilidae). *Misc. Zool. Hung.* 6:31–34.

Figure 4. Approximate neotropical distribution of *Balanteodrilus* and *Diplocardia*.

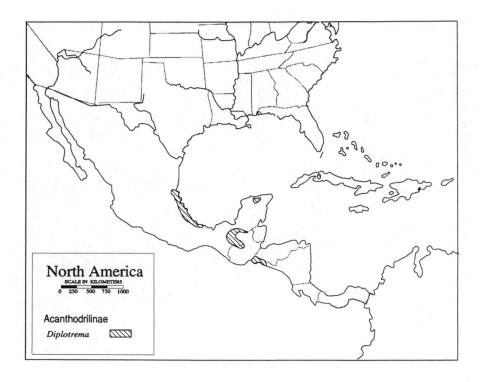

Figure 5. Approximate neotropical distribution of *Diplotrema*.

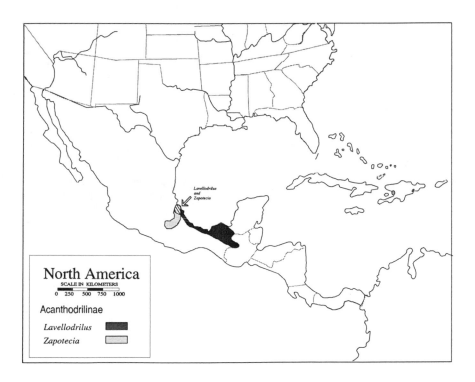

Figure 6. Approximate neotropical distribution of *Lavellodrilus* and *Zapotecia*.

Figure 7. Approximate neotropical distribution of *Zapatadrilus* and *Borgesia*.

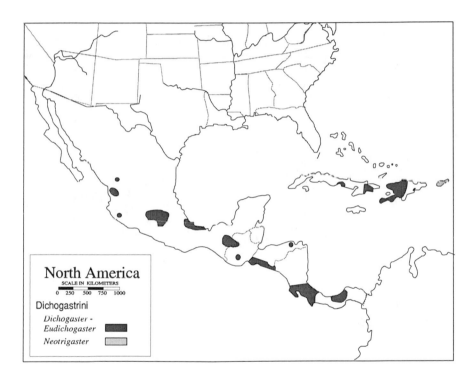

Figure 8. Approximate neotropical distribution of *Dichogaster-Eudichogaster* and *Neotrigaster*.

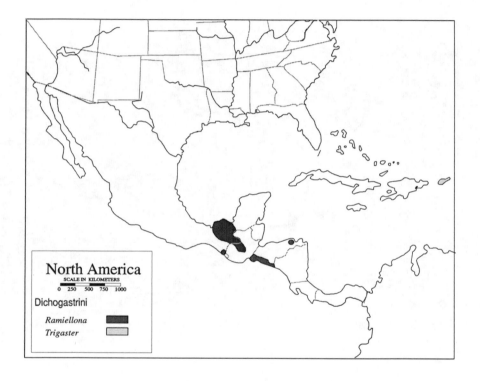

Figure 9. Approximate neotropical distribution of *Ramiellona* and *Trigaster*.

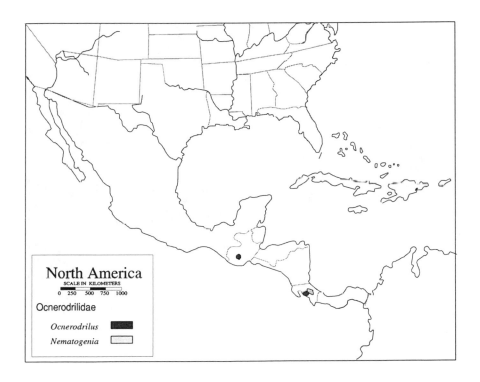

Figure 10. Approximate neotropical distribution of *Ocnerodrilus* and *Nematogenia*.

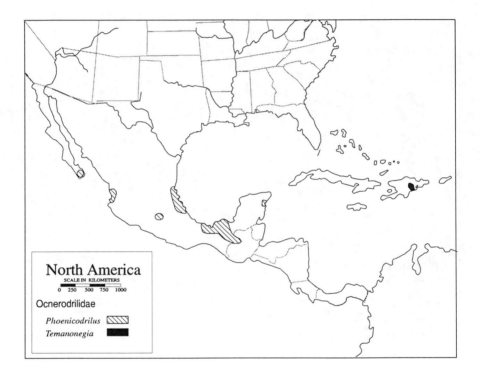

Figure 11. Approximate neotropical distribution of *Phoenicodrilus* and *Temanonegia*.

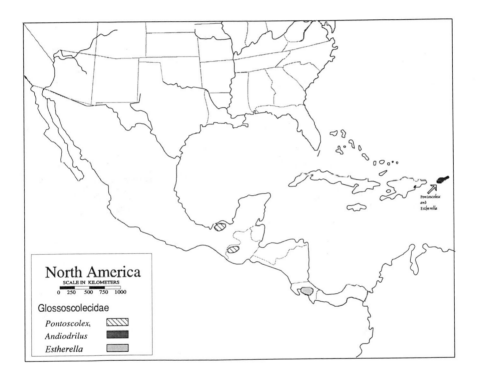

Figure 12. Approximate neotropical distribution of *Pontoscolex*, *Andiodrilus*, and *Estherella*.

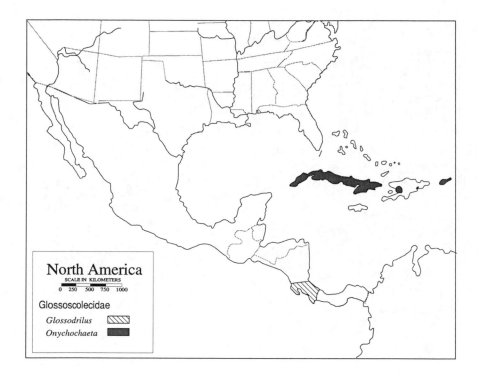

Figure 13. Approximate neotropical distribution of *Glossodrilus* and *Onychochaeta*.

Figure 14. Approximate neotropical distribution of *Mayadrilus* and *Protozapotecia*.

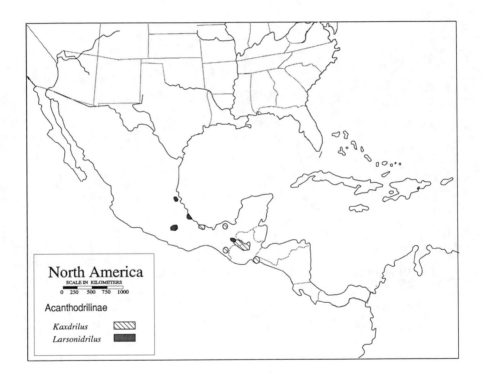

Figure 15. Approximate neotropical distribution of *Larsonidrilus* and *Kaxdrilus*.

Appendix 1. Native earthworms from Mexico, Central America, and the Greater Antilles.

	Country	Reference
MEGASCOLECIDAE		
Acanthodrilini		
Balanteodrilus pearsei	Mexico	Pickford 1938, Gates 1977a, Fragoso and Lavelle 1987, Fragoso 1990
Diplocardia eiseni	Mexico	Fragoso 1993
Diplocardia keyesi	Mexico	Eisen 1896
Diplocardia koebeli	Mexico	Eisen 1900
Diplotrema albidus	Mexico	Gates 1970b, 1973
Diplotrema haffneri	El Salvador	Graff 1957
Diplotrema jennifercie	Belize	Reynolds and Righi 1994
Diplotrema mexicana	Mexico	Gates 1967
Diplotrema murchie	Mexico	James 1990
Diplotrema oxcutzca-bensis	Mexico	Pickford 1938
Diplotrema papillata	Mexico	James 1990
Diplotrema ulrici	Cuba	Michaelsen 1923
Diplotrema vasliti	Mexico	Eisen 1896
Diplotrema whitmani	Guatemala	Eisen 1900
Diplotrema zilchi	Mexico	Fragoso 1993
	El Salvador	Graff 1957
Kaxdrilus crystallifer	Guatemala	Eisen 1990
Kaxdrilus hamiger	Guatemala	Michaelsen 1911
Kaxdrilus sylvicola	Mexico	Fragoso and Rojas 1994, Fragoso and Lavelle 1987
Kaxdrilus salvadorensis	El Salvador	Graff 1957
Kaxdrilus tamajusi	Guatemala	Eisen 1896
Kaxdrilus porcus	Mexico	Fragoso and Rojas 1994
Kaxdrilus proboscithecus	Mexico	Fragoso and Rojas 1994
Larsonidrilus microscolecinus	Mexico	James 1993
Larsonidrilus orbiculatus	Mexico	James 1993
Lavellodrilus bonampa-kensis	Mexico	Fragoso 1991
Lavellodrilus ilkus	Mexico	Fragoso 1991

Appendix 1. Native earthworms from Mexico, Central America, and the Greater Antilles. (continued)

	Country	Reference
Lavellodrilus maya	Mexico	Fragoso 1988
Lavellodrilus parvus	Mexico	Fragoso 1988
Lavellodrilus riparius	Mexico	Fragoso 1988
Mayadrilus calakmulensis	Mexico	Fragoso and Rojas 1994
Mayadrilus rombki	Mexico	Fragoso and Rojas 1994
Protozapotecia aquilonalis	Mexico	James 1993
Protozapotecia australis	Mexico	James 1993
Zapotecia amecameca	Mexico	Eisen 1900
Zapotecia keiteli	Haiti	Michaelsen 1902
Zapotecia nova	Mexico	James 1993
Dichogastrini		
Borgesia montana	Puerto Rico	James 1991
Borgesia sedecimsetae	Puerto Rico	Borges and Moreno 1991
Borgesia wegei	Puerto Rico	James 1991
Dichogaster greeffi	Virgin Islands	Michaelsen 1902
Dichogaster jamaicae	Jamaica	Eisen 1900
Dichogaster mexicana	Mexico	Michaelsen 1900
Dichogaster modigliani	Costa Rica	Righi and Fraile 1987
Dichogaster nana	Mexico	Eisen 1900
Dichogaster papillata	Mexico	Eisen 1896
Dichogaster pitahayana	Costa Rica	Michaelsen 1912
Dichogaster servi	Costa Rica	Righi and Fraile 1987
Dichogaster tristani	Costa Rica	Cognetti 1907
Eutrigaster (E) decorata	El Salvador	Graff 1957
Eutrigaster (G) eiseni	Mexico	Eisen 1900, Csuzdi and Ziczi 1991
Eutrigaster (E) franzi	Jamaica	Csuzdi and Ziczi 1991
Eutrigaster (G) gagzoi	Haiti	Michaelsen 1908
Eutrigaster (G) godeffroyi	Haiti	Michaelsen 1900
	Dominican Republic	Michaelsen 1936
Eutrigaster (E) grandis	Jamaica	Sims 1987
Eutrigaster (G) guatemala	Guatemala	Eisen 1900

Appendix 1. Native earthworms from Mexico, Central America, and the Greater
Antilles. (continued)

	Country	Reference
Eutrigaster (E) guetare	Costa Rica	Righi 1972b
Eutrigaster (G) hartme-yeri	Jamaica	Michaelsen 1908
Eutrigaster (E) hilaris	Costa Rica	Cognetti 1904a
Eutrigaster (G) keiteli	Haiti	Michaelsen 1900, 1902
Eutrigaster (E) kepo	Costa Rica	Righi and Fraile 1987
Eutrigaster (E) lineri	Mexico	Righi 1972b
Eutrigaster (G) manni	Haiti	Michaelsen 1935
Eutrigaster (G) michael-seniana	Costa Rica	Csuzdi and Ziczi 1991
Eutrigaster (E) montecya-nensis	Jamaica	Sims 1987
Eutrigaster (G) montana	Jamaica	Csuzdi and Ziczi 1991
Eutrigaster (E) oraedi-vitis	Costa Rica	Cognetti 1904b
Eutrigaster (E) orobia	El Salvador	Graff 1957
	Jamaica	Sims 1987
Eutrigaster (G) picadoi	Costa Rica	Michaelsen 1912
Eutrigaster (G) reichardti	Jamaica	Michaelsen 1935
	Haiti	
Eutrigaster (E) ribauco-urti	Mexico	Eisen 1900
Eutrigaster (G) sporado-nephra	Mexico	Lavelle and Kohlmann 1984
	El Salvador	Graff 1957
	Honduras	Righi 1972b
	Costa Rica	Cognetti 1908
Eutrigaster (E) townsendi	Jamaica	Eisen 1900
Eutrigaster (G) uhleri	Haiti	Michaelsen 1935
Eutrigaster (E) vialis	Guatemala	Michaelsen 1912
Eutrigaster (G) viridis	Mexico	Eisen 1900
Neotrigaster complutensis	Puerto Rico	Borges and Moreno 1991
Neotrigaster rufa	Puerto Rico	Gates 1954, 1962b
Neotrigaster yukiyui	Puerto Rico	Borges and Moreno 1991
Ramiellona americana	Guatemala	Gates 1962a
Ramiellona balantina	Guatemala	Gates 1962a

Appendix 1. Native earthworms from Mexico, Central America, and the Greater
Antilles. (continued)

	Country	Reference
Ramiellona eiseni	Guatemala	Michaelsen 1911
Ramiellona guatemalana	Guatemala	Gates 1962a
Ramiellona irpex	Mexico	Michaelsen 1923
	Guatemala	Michaelsen 1911
Ramiellona lasiura	El Salvador	Graff 1957
Ramiellona lavellei	Mexico	Gates 1978
Ramiellona* mexicana	Mexico	Gates 1962a
Ramiellona sauerlandti	El Salvador	Graff 1957
Ramiellona stadelmanni	Honduras	Michaelsen 1935
Ramiellona strigosa	Mexico	Fragoso and Lavelle 1987, Righi 1972b
	Guatemala	Gates 1962a
Ramiellona tecumumami	Guatemala	Michaelsen 1911
Ramiellona vulcanica	El Salvador	Graff 1957
Ramiellona* willsoni	Mexico	Righi 1972b
Trigaster calwoodi	Virgin Islands	James 1991, Michaelsen 1900
Trigaster intermedia	Virgin Islands	Michaelsen 1900
Trigaster lankesteri	Virgin Islands	Michaelsen 1900
Trigaster longissimus	Puerto Rico	Borges and Moreno 1991
Zapatadrilus albidus	Mexico	Gates 1973
Zapatadrilus cavernicolus	Cuba	Gates 1962b
Zapatadrilus reddeli	Mexico	Gates 1971
Zapatadrilus ticus	Mexico	Righi 1972b
Zapatadrilus toltecus	Mexico	Eisen 1900
Zapatadrilus vallesensis	Mexico	Gates 1971

OCNERODRILIDAE

Ocnerodrilinae

Nematogenia panamaensis	Costa Rica	Cognetti 1904b
Ocnerodrilus alox	Costa Rica	Righi and Fraile 1987
Ocnerodrilus tuberculatus	Guatemala	Eisen 1900
Phoenicodrilus taste	Mexico	Eisen 1895, Fragoso 1993
Temanonegia alba	Dominican Republic	Gates 1957, 1979
Temanonegia dominicana	Dominican Republic	Gates 1957, 1979

Appendix 1. Native earthworms from Mexico, Central America, and the Greater Antilles. (continued)

	Country	Reference
Temanonegia magna	Dominican Republic	Gates 1957, 1979
Temanonegia montana	Dominican Republic	Gates 1957, 1979
GLOSSOSCOLECIDAE		
Andiodrilus biolleyi	Costa Rica	Cognetti 1904a, Michaelsen 1912, Righi 1972b
Andiodrilus orosiensis	Costa Rica	Michaelsen 1912
Diachaeta thomasi	Jamaica	Michaelsen 1900
	Virgin Islands	Michaelsen 1900
Estherella aguayoi	Puerto Rico	Borges and Moreno 1989
Estherella caudoferruginea	Puerto Rico	Borges and Moreno 1989
Estherella gatesi	Puerto Rico	Borges and Moreno 1989
Estherella montana	Puerto Rico	Gates 1970a
Estherella nemoralis	Puerto Rico	Gates 1970a
Estherella stuarti	Puerto Rico	Borges and Moreno 1989
Estherella toronegrensis	Puerto Rico	Borges and Moreno 1989
Glossodrilus nemoralis	Costa Rica	Cognetti 1905a, 1906
Glossodrilus orosi	Costa Rica	Righi and Fraile 1987
Glossodrilus dorasque	Costa Rica	Righi and Fraile 1987
Glossodrilus cibca	Costa Rica	Righi and Fraile 1987
Onychochaeta elegans	Cuba	Michaelsen 1923
Onychochaeta windlei	Cuba	Michaelsen 1923
	Haiti	Michaelsen 1935
	Virgin Islands	Righi 1971
Pontoscolex cynthiae	Mexico	Fragoso 1993
	Puerto Rico	Borges and Moreno 1990b
Pontoscolex lilljeborgi	Guatemala	Eisen 1896
Pontoscolex melissae	Puerto Rico	Borges and Moreno 1990b
Pontoscolex spiralis	Puerto Rico	Borges and Moreno 1990b

* Species that will be placed in a different, new genus (Fragoso and Rojas, unpublished data)

Native and Exotic Earthworms in Wildland Ecosystems

Paul J. Kalisz and Hulton B. Wood

I.   Introduction

A.   Objectives

Our objective was to summarize current knowledge concerning the patterns of occurrence of native and exotic earthworms (Annelida; Oligochaeta) in wildland ecosystems, and to discuss documented and hypothetical types of interactions that could produce the observed patterns. To do this we interpreted data obtained during our individual research projects in southern California (HBW) and Kentucky (PJK) in terms of local land-use history and earthworm biology. We hope our interpretation provides a preliminary framework for understanding native-exotic earthworm interactions in wildland ecosystems in other parts of North America and the world. We also hope our discussion and speculations stimulate research on the various possible mechanisms of interaction between native and exotic earthworms, and on the relationship between habitat disturbance and dominance by exotic earthworm species.

1-56670-053-1/95/$0.00+$.50
©1995 by CRC Press, Inc.

We defined wildland ecosystems as those that are occupied chiefly by native plants and animals; not intensively used as urban or residential areas; and not intensively managed for the production of domesticated plants or animals. We considered a wildland to be "fragmented" when most or all of the original vegetation had been removed, and native vegetation occurred only in small patches or "remnants." Our definition of "slight" disturbance allowed logging but not land clearance, or conversion to agricultural fields, pastures, urbanized areas, and other types of intensive human use.

B.   Native and Exotic Earthworm Taxa

With only minor differences among taxonomists, four families and seven genera of "native" earthworms that we recorded in our study areas (Table 1) are considered endemic to North America (Gates 1982, Sims and Gerard 1985, Fender and McKey-Fender 1990, James 1990, Schwert 1990). There is less complete agreement (see discussion in Kalisz 1993) as to whether all of the "exotic" genera that we recorded (Table 2) are truly exotic in the sense that they were absent from North America 400 to 500 years ago at the time of arrival of European colonists. In any case, our "exotic" taxa were apparently uncommon in undisturbed wildlands south of the glacial limits as recently as 15 to 30 years ago (e.g., Stebbings 1962, Gates 1970, Reynolds 1974, Reynolds et al. 1974). We therefore considered as exotic invaders of wildland ecosystems, 11 recorded taxa of earthworms (Table 2) that occur naturally in other parts of the world but presently dominate portions of North America that have been intensively disturbed and portions that were covered by ice during the Wisconsinan glaciation.

Table 1. Native families and genera of earthworms recorded in southern California (CA) and eastern Kentucky (KY). Taxonomy according to James (1990) and Fender and McKey-Fender (1990).

KOMAREKIONIDAE	*Komarekiona* Gates, 1974 (KY)
LUMBRICIDAE	*Bimastos* Moore, 1893 (KY)
	Eisenoides Gates, 1969 (KY)
MEGASCOLECIDAE	*Argilophilus* Eisen, 1893 (CA)
	Diplocardia Garman, 1888 (CA, KY)
OCNERADRILIDAE	*Ocnerodrilus* Eisen, 1878 (CA)
SPARGANOPHILIDAE	*Sparganophilus* Benham, 1892 (CA)

Table 2. Exotic families and genera of earthworms recorded in southern California (CA) and eastern Kentucky (KY). Taxonomy according to James (1990), Fender and McKey-Fender (1990), and Schwert (1990).

LUMBRICIDAE	*Allolobophora* Eisen, 1874 (CA, KY)
	Aporrectodea Örley, 1885 (CA, KY)
	Dendrobaena Eisen, 1873 (CA)
	Dendrodrilus Omodeo, 1956 (KY)
	Eisenia Malm, 1877 (CA, KY)
	Eiseniella Michaelsen, 1900 (CA)
	Lumbricus L., 1758 (CA, KY)
	Murchieona Gates, 1978 (KY)
	Octolasion Örley, 1885 (CA, KY)
MEGASCOLECIDAE	*Microscolex* Rosa, 1887 (CA)
	Pheretima Kingberg, 1867 *sensu lato* (KY)

C. Native-Exotic Earthworm Interactions

The potential for extirpation of populations and species of native earthworms due to habitat destruction and the introduction of earthworms from Europe and Asia has been a concern for a long time (Eisen 1900, Smith 1928, Ljungstrom 1972, Gates 1977). About 30 years ago Stebbings (1962), working in Missouri, concluded that exotic earthworms were replacing native earthworms in situations where forest ecosystems were severely disturbed by human activities. Research in wildland ecosystems in New Zealand (Miller et al. 1955, Lee 1961), Australia (Wood 1974, Abbott 1985), South Africa (Ljungstrom and Reinecke 1969), and the United States (Kalisz and Dotson 1989) supports the following hypothetical sequence of events leading to dominance by exotic earthworms: (1) severe disturbance or destruction of the habitat; (2) reduction in size or extirpation of native earthworm populations; (3) introduction of exotic species of earthworms; and (4) colonization of empty habitat by exotic earthworms. As discussed later in this paper, in addition to disturbance-induced replacement of native earthworms, there are other possible types of interactions between native and exotic earthworms. For example, native earthworms may be displaced by exotics through direct competition, or exotic earthworms may colonize niches not occupied by natives and coexist with native taxa.

II.   Summary of Research in Kentucky and Southern California

A.   Introduction

Our generalizations and interpretations are based on data from studies in Kentucky and southern California that dealt specifically with earthworms, and on notes and insights collected in the course of these and other research projects. These studies sampled a variety of severely and slightly disturbed habitats located in both non-fragmented and fragmented wildlands. Research in Kentucky was performed in remnants of the fragmented forests of the Bluegrass physiographic region, and in the extensive and non-fragmented forests of the Cumberland Plateau. Sampling in southern California was performed in riparian, woodland, grassland, and chaparral habitats in 10 geographic areas selected to represent the Mediterranean climatic zones of southern California. Detailed methods and results are given in Dotson and Kalisz (1989), Kalisz and Dotson (1989), Kalisz (1993), and Wood and James (1993).

B.   Kentucky

In general, undisturbed or slightly disturbed sites in the non-fragmented forests of the Cumberland Plateau were occupied only by native earthworms; exotic taxa occurred only in areas that had been severely disturbed. As documented previously (Kalisz and Dotson 1989, Kalisz 1993), native earthworm taxa showed variable sensitivity to disturbance: the genus *Diplocardia* Garmon was relatively insensitive and persisted even on some cleared and cultivated sites, whereas *Komarekiona eatoni* Gates was very sensitive and occurred only on minimally disturbed sites. Seven native and 5 exotic taxa were recorded on 57 sites. All-native assemblages were found on 49 slightly disturbed sites; an all-exotic earthworm assemblage was found on a single severely disturbed site; and mixed assemblages were found on 7 severely disturbed sites.

The relationship between the occurrence of exotic earthworms and disturbance was different in the fragmented wildlands of the Bluegrass region. Exotic species dominated regardless of the size of the forest remnant and the intensity of disturbance, occurring on all severely disturbed sites and on 18 of the 21 slightly disturbed sites. Nine exotic and 5 native taxa were recorded in forest remnants in the Bluegrass. *Diplocardia* was the most common native taxon. This endemic genus was recorded from 2 of the 3 sites that had all native earthworms, and from all 3 of the sites that had both native and exotic earthworms.

C.   Southern California

Riparian habitats supported the greatest diversity and abundance of earthworms, and were in many cases entirely occupied by exotic lumbricids; no riparian sample sites were found to contain an all-native earthworm assemblage. In aquatic or water-logged sites the native genera *Sparganophilus* Benham and *Ocnerodrilus* typically were found in association with introduced lumbricids.

In oak (*Quercus* L.) woodlands, soils in the influence zone of oak trees were typically dominated by exotic earthworm taxa even in cases where there was no evidence of past disturbance to the site. Grassy areas between trees contained both native and exotic earthworms, but were dominated by the native genera *Argilophilus* Eisen or *Diplocardia*.

Grasslands composed entirely of native species of grass are rare in southern California, and only one such grassland was sampled. This area was occupied by a single species, the native earthworm *Argilophilus papillifer* Eisen. Grasslands composed of non-native grasses, on the other hand, were common and were occupied by earthworm assemblages composed of both exotic and native taxa, including the native genera *Argilophilus* and *Diplocardia*. Exotic taxa were clearly dominant on disturbed areas such as abandoned house sites, watering ponds, creek channels, and plantations of exotic trees, but became less completely dominant with increasing distance away from such disturbance centers.

Steep, shallow soils under chaparral were characterized by very sparse populations of exotic taxa. The distribution of earthworms was patchy, and earthworms were limited to the litter-covered soil areas beneath individual chaparral plants. No native earthworms were found on these poor chaparral sites, suggesting that exotic taxa were in this case occupying a habitat that was not utilized by native earthworms.

III.   Discussion and Speculation

Our conclusion is that the occurrence and relative dominance of native and exotic earthworm taxa are determined by the disturbance history of both the sampling site and the landscape containing the sampling site, and by the habits and adaptability of the taxa that interact. In nonfragmented landscapes Elton's (1958:154) observation that exotic invasions occur most readily in "changed and simplified" habitats seems to hold, and replacement of native earthworm assemblages seems to only follow severe habitat disturbance. When dealing with forest remnants in disturbed landscapes, however, replacement of native taxa seems to occur even when the forest remnants appear undisturbed and large enough to support viable populations of native earthworms. This discrepancy is not surprising: although soil volumes and aboveground areas larger than the effective universe of individual earthworms may remain intact, remnants differ from nonfragmented wildlands in many ways including plant and animal species diversity and abundance; microclimate; extent of "edge"; influx of propagules or individuals of exotic taxa; rates of extinction; competition and predation among species; and inputs of mass and energy (Wilcox and Murphy 1985, Klein 1989, Laurance and Yensen 1991, Saunders et al. 1991), possibly including human inputs of toxic chemicals. Although many of these results of habitat fragmentation are invisible, they apparently reduce the ability of native earthworms to survive or to compete with introduced species.

Most earthworm introductions have resulted from human transport of soils and plants (Gates 1982, Lee 1985). In North America, human population expansions and shifts over the centuries have likely been responsible for the re-distribution of some native earthworm taxa (e.g., the genus *Bimastos* Moore) and for the introduction and spread of exotic earthworm taxa. Exotic earthworms often become established along lakes and streams because of discarded fish bait (e.g., *Lumbricus castaneus* Savigny at site J of Kalisz and Dotson 1989), and floods are known to carry viable cocoons and live earthworms. For these reasons, earthworm assemblages in riparian zones and stream bottoms often consist wholly or partly of exotic taxa, and many seemingly pristine areas along waterways are in reality severely disturbed in terms of species introductions and native-exotic species interactions. In recent years, the planting of trees and the cultivation of marijuana (*Cannabis sativa* L.) in remote areas may function as mechanisms for local introductions of exotic earthworms. In addition to humans, birds and other animals are considered likely dispersal agents (Schwert 1980).

In some cases, the occurrence of exotic taxa, and the chronology and mechanisms of invasions, seem to be unpredictable and extremely fortuitous. The occurrence of *Eiseniella tetraedra* Savigny and two other exotic species in a remote and uninhabited canyon on the San Dimas Experimental Forest in southern California (Wood and James 1993), and the occurrence of a single small population of *Lumbricus terrestris* L. in a remote hollow on the Robinson Experimental Forest in Kentucky (Kalisz and Dotson 1989) are examples of exotic invasions that are not easily explained in terms of habitat disturbance or land-use history.

We conclude that: (1) in fragmented wildlands native earthworms will likely disappear even from relatively undisturbed remnants; (2) in nonfragmented wildlands native earthworms will disappear only in the immediate vicinity (within ~50 m) of severe disturbance. In the case of nonfragmented habitat the sequence of events is likely as documented in the literature: habitat destruction leads to the extirpation of local populations of native earthworms, then the vacant habitat is occupied by introduced species. Based on this paradigm, direct competition between native and exotic species is not essential to the replacement process. In fragmented habitat, however, visible habitat disturbance does not seem to be an essential step in the replacement process. This means that: (1) exotic species directly out-compete and displace native species in relatively intact remnants of the original wildland ecosystem; or (2) invisible but ecologically significant changes in earthworm habitat result from fragmentation and lead to extirpation of native taxa. The latter alternative, that "invisible" disturbance rather than alteration of the physical structure of the habitat initiates the replacement process, ultimately leads to dominance by exotics via the same sequence of events described for the case of severe disturbance in non-fragmented landscapes. Direct competition between native and exotic earthworms has not been documented, but could conceivably occur in fragmented habitat due to the large population of exotic earthworms in the encompassing disturbed landscape and to the greatly increased perimeter ("edge") across which exotic species could invade the remnant. Circumstantial evidence of direct displacement of native by exotic earthworms was provided by the observation in southern California that exotic taxa dominated the most favorable habitats of riparian zones and native taxa commonly occurred only on the fringes, suggesting that the natives had been displaced outward to drier areas by introduced earthworms. Better understanding of species life-histories, the habits and requirements of native and exotic earthworm taxa, and the structure of earthworm communities (*sensu* Lavelle 1983) is required to determine the likelihood of niche-overlap and direct competition between native and exotic taxa.

Regardless of the exact mechanism of native-exotic earthworm interaction, we do not feel that native earthworms will survive over long periods in wildland remnants in fragmented landscapes, especially in remnants <30 ha in area. In both Southern California and Kentucky the endemic genus *Diplocardia* may be an exception to this generalization. Since members of this genus typically live in the subsoil and consume humus, they may be insensitive to many types of disturbance that adversely affect species that live near to the surface and may be much less sensitive to changes in vegetation than species that consume plant parts or freshly-fallen litter. Deep-living endemic earthworm taxa, ecologically similar to *Diplocardia*, have been reported to survive in severely disturbed areas in New Zealand (Miller et al. 1955), South Africa (Ljungstrom and Reinecke 1969), and Australia (Wood 1974). Although exotic earthworm taxa may ultimately dominate fragmented wildland landscapes, in some cases well-adapted endemic species may persist on sites that are droughty, steep or inaccessible due to both the hostility of such sites and to the low frequency and intensity of human disturbance and import of exotic species.

Finally, we propose that earthworm assemblages integrate and reflect conditions over a relatively large area which encompasses and affects their living space. This means that the occurrence and dominance of native earthworms may be useful as an index of ecosystem "integrity" on the scale of the sampling site in nonfragmented landscapes, but would need to be interpreted on the scale of the landscape, rather than of the sampling site, when evaluating wildland remnants in fragmented landscapes.

IV.   Research Imperatives

We recommend that the following general types of research be initiated to address questions regarding mechanisms of interaction between native and exotic earthworm species.

A.   Natural Experiments

These would involve identification of and long-term monitoring across gradual or abrupt contacts between earthworm assemblages dominated by native and exotic taxa. Such monitoring would document spatial and temporal changes in species dominance, especially changes that are too slow to be recognized during the short observation periods typical of most research projects.

B.   Controlled Experiments

These would involve introductions or alteration of the dominance of earthworm species in captivity or in the field. Introductions of single species or of naturally-occurring species assemblages (native, exotic, or mixed), with and without habitat disturbance, could be used to examine the importance of direct competition versus habitat disturbance in the shift from native-dominated to exotic-dominated earthworm assemblages.

C.   Surveys

These would document existing patterns of occurrence and dominance of native and exotic earthworm taxa in interesting or relatively unknown wildlands. We recommend that the full range of locally-occurring undisturbed and human-influenced ecosystems be sampled in such surveys, and that sufficient information on local land-use history, climate, and soils be obtained to allow interpretation of the survey results.

References

Abbott, I. 1985. Distribution of introduced earthworms in the northern jarrah forest of western Australia. *Aust. J. Soil Res*. 23:263–270.
Dotson, D.B. and P.J. Kalisz. 1989. Characteristics and ecological relationships of earthworm assemblages in undisturbed forest soils in the southern Appalachians of Kentucky, USA. *Pedobiologia* 33:211–220.

Eisen, G. 1900. Researches in American Oligochaeta, with especial reference to those of the Pacific Coast and adjacent islands. *Proc. Calif. Acad. Sci.*, Third Series, II(2):85– 276.

Elton, C.S. 1958. *The ecology of invasions by animals and plants*. Chapman and Hall, London. 181 pp.

Fender, W.M. and D. McKey-Fender. 1990. Oligochaeta: Megascolecidae and other earthworms from western North America. p. 357–378. In: Dindal, D.L. (ed.) *Soil biology guide*. John Wiley and Sons, New York.

Gates, G.E. 1970. Miscellanea megadrilogica. VIII. *Megadrilogica* 1:1–14.

Gates, G.E. 1977. On the correct generic name for some West Coast native earthworms, with aids for a study of the genus. *Megadrilogica* 3:54–60.

Gates, G.E. 1982. Farewell to North American megadriles. *Megadrilogica* 4:12–77.

James, S.W. 1990. Oligochaeta: Megascolecidae and other earthworms from southern and midwestern North America. p. 379–386. In: Dindal, D.L. (ed.) *Soil biology guide*. John Wiley and Sons, New York.

Kalisz, P.J. 1993. Native and exotic earthworms in deciduous forest soils of eastern North America. p. 93–100. In: McKnight, B.N. (ed.) *Biological pollution: the control and impact of invasive exotic species*. Indiana Academy of Sciences, Indianapolis.

Kalisz, P.J. and D.B. Dotson. 1989. Land-use history and the occurrence of exotic earthworms in the mountains of eastern Kentucky. *Am. Midl. Nat.* 122:288–297.

Klein, B.C. 1989. Effects of forest fragmentation on dung and carrion beetle communities in central Amazonia. *Ecology* 70:1715–1725.

Laurance, W.F. and E. Yensen. 1991. Predicting the impacts of edge effects in fragmented habitats. *Biol. Conserv.* 55:77–92.

Lavelle, P. 1983. The structure of earthworm communities. p. 449–466 In: Satchell, J.E. (ed.) *Earthworm ecology from Darwin to vermiculture*. Chapman and Hall, London.

Lee, K.E. 1961. Interactions between native and introduced earthworms. *Proc. N.Z. Ecol. Soc.* 8:60–62.

Lee, K.E. 1985. *Earthworms: their ecology and relationships with soils and land use*. Academic Press, New York.

Ljungstrom, P.-O. 1972. Introduced earthworms of South Africa. On their taxonomy, distribution, history of introduction and on the extermination of endemic earthworms. *Zool. Jahrb. Syst. Bd.* 99:1–81.

Ljungstrom, L.-O. and A.J. Reinecke. 1969. Ecology and natural history of the microchaetid earthworms of South Africa. 4. Studies on influence of earthworms upon the soil and parasitological question. *Pedobiologia* 9:152–157.

Miller, R.B., J.D. Stout, and K.E. Lee. 1955. Biological and chemical changes following scrub burning on a New Zealand hill soil. *N.Z. J. Sci. Tech.* 37:290–313.

Reynolds, J.W. 1974. The earthworms of Maryland (Oligochaeta: Acanthrodrilidae, Lumbricidae, Megascolecidae and Sparganophilidae). *Megadrilogica* 1:1–12.

Reynolds, J.W., E.E.C. Clebsch, and W.M. Reynolds. 1974. Contributions to North American earthworms. The earthworms of Tennessee (Oligochaeta). I. Lumbricidae. *Bull. Tall Timbers Res. Stn.* 17:1–133.

Saunders, D.A., R.J. Hobbs, and C.R. Margules. 1991. Biological consequences of ecosystem fragmentation: a review. *Conserv. Biol.* 5:18–32.

Schwert, D.P. 1980. Active and passive dispersal of Lumbricid earthworms. p. 182–189 In: Dindal, D.L. (ed.) *Soil biology as related to land use practices*, Proc. 7th Inter. Colloquium Soil Zoology. Environmental Protection Agency, Washington, D.C.

Schwert, D.P. 1990. Oligochaeta: Lumbricidae. p. 341–356 In: Dindal, D.L. (ed.) *Soil biology guide*. John Wiley and Sons, New York.

Sims, R.W. and B.M. Gerard. 1985. *Earthworms*. E.J. Brill, London. 171 pp.

Smith, F. 1928. An account of changes in the earthworm fauna of Illinois and a description of one new species. *Ill. Nat. His. Surv. Bull.* 17:347–362.

Stebbings, J.H. 1962. Endemic-exotic earthworm competition in the American Midwest. *Nature* 196:905–906.

Wilcox, B.A. and D.D. Murphy. 1985. Conservation strategy: the effects of fragmentation on extinction. *Am. Nat.* 125:879–887.

Wood, H.B. and S.W. James. 1993. Native and introduced earthworms from selected chaparral, woodland, and riparian zones in southern California. *USDA Forest Service Gen. Tech. Rep.* PSW-GTR-142.

Wood, T.G. 1974. The distribution of earthworms (Megascolecidae) in relation to soils, vegetation and altitude on the slopes of Mt. Kosciusko, Australia. *J. Animal Ecol.* 43:87–106.

Influences of Earthworms on Biogeochemistry

John M. Blair, Robert W. Parmelee, and Patrick Lavelle

1-56670-053-1/95/$0.00+$.50
©1995 by CRC Press, Inc.

127

I.   Introduction

The importance of earthworms in affecting soil structure, organic matter processing and nutrient cycling has long been recognized (Darwin 1881, Edwards and Lofty 1977, Lee 1985). However, there are many areas of earthworm biology and ecology which are insufficiently studied, and the overall effects of these important invertebrates on decomposition and nutrient cycling are still poorly understood (Lavelle 1988). This is particularly true in North America, where soil ecology only recently has begun to receive the degree of attention it has been afforded in Europe, New Zealand, and Australia. Available data on specific aspects of earthworm biology and ecology indicate the potential of earthworms to significantly influence soil structure; organic matter comminution, distribution, and chemistry; soil microbial community structure and activity; soil nutrient transformations (nutrient mineralization/immobilization, N fixation, nitrification, denitrification); plant uptake of nutrients; and potential losses in runoff and soil leachate (see reviews in Edwards and Lofty 1977, Satchell 1983, Lee 1985 and references in the following sections). Additionally, the storage and turnover of nutrients in earthworm biomass is significant in many terrestrial ecosystems. However, in spite of considerable component research on earthworm biology and ecology, critical questions remain unanswered about the effects of earthworms on ecosystem-level processes in both natural ecosystems and agroecosystems in North America.

The present state of knowledge regarding the effects of earthworms on biogeochemical cycling in North American ecosystems is based largely on fragmentary studies done at different locations with different species, and by extrapolating results from other regions of the world. This is due, in part, to the relatively small number of earthworm researchers in North America. Furthermore, there are complicating factors introduced by the effects of glaciation on the present distribution of earthworms, the displacement of native North American earthworm species by introduced species (see Kalisz and Wood, Chapter 5, this volume), and the effects of land use practices on soil communities and processes. With a few exceptions (e.g., Spiers et al. 1986, Parmelee and Crossley 1988, James 1991), the lack of intensive studies of the effects of earthworms in specific ecosystem types in North America is notable, as are the paucity of studies dealing specifically with native North American earthworm species. Within these constraints, we have attempted to construct a general picture of the mechanisms by which earthworms influence the biogeochemistry of North American ecosystems. We have drawn on results of studies done in North America, where possible, and made some inferences about the potential effects of earthworms on nutrient cycling processes based on studies done elsewhere. Our hope is that this approach will point out some areas where earthworms are likely to exhibit significant effects on biogeochemical cycles, and some areas that are in need of further study.

II. Ecological Groupings of Earthworm Species

We preface our discussion by pointing out that the effects of earthworms on biogeochemical processes can vary with different species of earthworm as well as with different soil types, organic matter resources, climatic regimes, and so on (Shaw and Pawluk 1986a, b, Lavelle 1988, Wolters and Joergensen 1992). These differences become important when comparing results from different studies and when extrapolating results from one ecosystem, or region, to another. Three major ecological groupings of earthworms have been defined, based primarily on feeding and burrowing strategies (Bouché 1977). Epigeic species live in or near the surface litter, and feed primarily on coarse particulate organic matter. They are typically small, and have high metabolic and reproductive rates as adaptations to the highly variable environmental conditions at the soil surface. Endogeic species live within the soil profile and feed primarily on soil and associated organic matter (geophages). They generally inhabit temporary burrow systems which are filled with cast material as the earthworms move through the soil. There is evidence that some endogeic species do not feed indiscriminately, but may preferentially ingest soil high in organic matter (James and Cunningham 1989, Judas 1992). Anecic species live in more or less permanent vertical burrow systems which may extend several meters into the soil profile. They feed primarily on surface litter which they pull into their burrows. Some anecic species also may create "middens" at the burrow entrance, consisting of a mixture of cast soil and partially incorporated surface litter.

The habitats and feeding preferences of different earthworm species relate to their potential effects on biogeochemistry. Anecic species, such as *Lumbricus terrestris*, may significantly accelerate the disappearance and breakdown of surface litter, and the macropores created by their burrows may be important in moving water and solutes into lower soil horizons. Their casting activity also may be important in soil profile formation. Geophagous earthworms feed in the soil, but can interact synergistically with anecics. For example, Shaw and Pawluk (1986a) found that *L. terrestris* played a dominant role in controlling decomposition rates in different soil types by mixing surface litter and soil, but this effect was attenuated when geophagous species were included in the soil incubations. This points out the need to consider the effects of entire earthworm communities when evaluating their impact on ecosystem processes. We should also indicate that the ecological categories discussed above are very broad and may not apply equally well to temperate and tropical ecosystems. Refinement of these classifications and questions of functional equivalency of earthworm species in various ecosystems remain to be addressed.

III. Influences on Carbon and Nutrient Fluxes at Different Spatial and Temporal Scales

A better understanding of earthworm influences on biogeochemistry of ecosystems will require studies at a range of spatial and temporal scales. There is ample evidence that effects of earthworms on nutrient cycling processes are highly scale-dependent, and apparently contradictory results may be observed at different levels of spatial and temporal resolution. Therefore, the questions asked and the interpretation of results must explicitly take into consideration the spatiotemporal scales at which the effects of earthworms are measured, including their influences on soil physicochemical properties and their interactions with other soil biota.

Lavelle and Martin (1992) discuss four scales at which effects of earthworms on soil organic matter dynamics are expressed in soils of the humid tropics: (1) gut transit; (2) fresh soil casts; (3) aging casts in the soil; and (4) long-term evolution of the whole soil profile. We believe it would be useful to consider similar scales, and the interrelationships among them, when assessing the potential effects of earthworms on organic matter and nutrient dynamics in temperate, subtropical, and tropical North American ecosystems. However, we need to point out that some important differences exist between the feeding behavior and ecology of tropical and temperate earthworms which may have consequences for how earthworms affect soil formation and processes. The extent to which results from tropical studies can be extrapolated to temperate North American ecosystems is unclear.

An important obstacle to overcome in assessing the net effects of earthworm activity on ecosystem processes involves scaling up from the microsite to the ecosystem level. How are effects measured at the microsite level (casts, burrow walls) related to effects at the whole soil, ecosystem, or landscape level? Our discussion of earthworm influences on biogeochemical processes begins by considering storage and turnover of nutrients in earthworm biomass, a direct effect. We then consider shorter-term effects of earthworm interactions with soils, organic matter and other biota at the microsite level, and discuss potential longer-term effects of earthworm activity at the whole soil level. We conclude with some recommendations for future research on earthworms and biogeochemistry in North America.

IV. Storage and Turnover of Nutrients in Earthworm Biomass

Earthworms play a direct role in the biogeochemistry of many terrestrial ecosystems as carbon and nutrients are accumulated, stored and cycled through earthworm biomass (ingestion, assimilation, respiration/excretion, mortality). Although earthworms can ingest large quantities of organic matter annually

(Edwards and Lofty 1977, Lee 1985), it generally is assumed that their direct contribution to total heterotrophic respiration is small. While earthworm respiration is often the major portion of faunal respiration, it usually is only 5–6% of the total energy flow in terrestrial ecosystems (Lee 1985). The low contribution of earthworms to C flux is due partly to their low C assimilation efficiencies. Estimates of the assimilation efficiencies of endogeic earthworms range from 2–6% to a maximum of 18% (Bolton and Phillipson 1976, Barois et al. 1987, Martin et al. 1992a). Reported assimilation efficiencies of litter-feeding earthworms are highly variable. For example, assimilation of leaf litter by *Lumbricus rubellus* was estimated to be 30–70% depending on the temperature and quality of the litter (Dickschen and Topp 1987). However, Crossley et al. (1971), using a cesium tracer, reported much lower assimilation efficiencies (12–29%) for a number of other earthworm species.

In some ecosystems with high earthworm biomass, and presumably activity, earthworm respiration can represent a significant C flux. For example, Parmelee et al. (1990), in a no-tillage agroecosystem in the southeastern U.S., observed maximum earthworm densities approaching 1000 ind. m^{-2} with a maximum biomass near 30 g ash-free dry mass m^{-2}. Hendrix et al. (1987) constructed a C budget for this system and estimated that during the cool and wet winter-spring season earthworms were responsible for about 30% of total heterotrophic respiration. More field-based estimates of earthworm assimilation, respiration and production in a variety of North American ecosystems are needed to better discern the direct contributions of earthworms to ecosystem carbon flux.

Unlike carbon, it appears that significant amounts of nitrogen move through earthworm biomass in many terrestrial ecosystems. In early work by Satchell (1963), N returned to the soil in dead *L. terrestris* tissue was estimated to be 60–70 kg N ha^{-1} yr^{-1}. When N returned in urine and mucus was included, estimates of total N flux approached 100 kg N ha^{-1} yr^{-1}. Other estimates from New Zealand forests and Polish pastures (reviewed in Lee 1985) also indicate a significant flux of N through earthworm biomass. In North America, Parmelee and Crossley (1988) estimated secondary productivity of the earthworm community in a Georgia no-tillage agroecosystem, and calculated the total N flux through earthworms to be over 60 kg N ha^{-1} yr^{-1}. This is a significant flux of N in this system, equivalent to about 38% of N uptake by the crop. Turnover of N (and C) in earthworm tissue appears to be rapid. Ferriere and Bouché (1985) labelled the anecic earthworm *Nicodrilus longus* by feeding it double labelled (^{14}C and ^{15}N) algae, and determined that the entire C and N content of the earthworm could turnover in as little as 40 days. Barois et al. (1987) incorporated ^{15}N into the tissue of *Pontoscolex corethrurus*, a tropical geophage, and found that 14% of the incorporated ^{15}N was eliminated within 5 days, and 30% was eliminated and replaced within 30 days.

Although the large amount of N moving through earthworm biomass in many systems is striking, it may be of greater significance that earthworm tissue, mucus, and urine are labile and easily assimilable N forms. Satchell (1967) estimated that a minimum of 70% of the N in dead earthworm tissue was mineralized in 10–20 days. Christensen (1988) also noted rapid mineralization from dead earthworms and calculated that 29–42 kg N ha^{-1} was released from dead earthworm tissue during the autumn in a Danish agroecosystem. Nitrogen released from dead earthworms can result in rapid recycling of N on relatively short time scales (Lee 1985). We are unaware of studies where direct fluxes of elements other than C and N through earthworm biomass have been investigated.

In addition to direct turnover of nutrients though their biomass, earthworms can influence biogeochemical processes in a variety of indirect ways. These indirect effects primarily involve interactions of earthworms with soil structure, organic resources, and other soil biota, and have important consequences for both short-term nutrient fluxes and long-term ecosystem dynamics.

V.   Effects of Earthworms at the Microsite Level

Many of the most readily measurable, and best documented, effects of earthworms on biogeochemical processes occur at the microsite level. These localized areas which exhibit marked effects of earthworm activity include casts, burrows and middens. Defining the precise area of influence for these microsites can be problematic. While casts and middens may be fairly discrete units, the distance to which effects of earthworm activity extend away from these sites and from burrow walls can be variable. Bouché (1975) defined the drilosphere as the 2 mm-thick zone around the walls of an earthworm burrow. However, the actual area of influence may extend more or less than this depending on the factor being considered. The rhizosphere also is an important site for earthworm-microbe-plant interactions. Most of the effects of earthworms at the microsite scale are a result of combined direct and indirect effects. For example, earthworms may increase N availability in soils directly, by adding excreted N to soil as it passes through the gut or through pores open to the external soil environment, and indirectly by stimulating microbial mineralization of organic soil N in cast material or in the linings of earthworm burrows, or by increasing rates of N fixation.

A.   Earthworm Casts

Many studies have demonstrated that earthworm casts have different chemical, biological, and physical characteristics than the bulk soil from which the casts were produced, although the results are sometimes contradictory (Parle 1963, Dkhar and Mishra 1986, Scheu 1987, Wolters and Joergensen 1992). In fact, the specific effects reported in various studies seem to depend on many factors such as earthworm species (Shaw and Pawluk 1986a), soil type and texture (Wolters and Joergensen 1992), soil moisture (Mackay and Kladivko 1985), availability and quality of organic matter (Lavelle et al. 1980, Shipitalo and Protz 1988), and age of the casts at sampling (Scheu 1987, Wolters and

Joergensen 1992). Some studies compare field-collected casts of unspecified age, or broad age intervals, to the surrounding bulk soil (i.e., Tiwari et al. 1989, Dkhar and Mishra 1986, Spiers et al. 1986, Elliott et al. 1990). Others have compared casts produced from homogenized soils in laboratory incubations to control soils without earthworms (i.e., Svensson et al. 1986). Sometimes the casts from these incubations are of a specific age when collected (Scheu 1987, Lavelle et al. 1992) and sometimes they are of mixed ages (Shaw and Pawluk 1986a). Yet other studies have compared the whole soils in which earthworms have been incubated (casts, burrow linings, bulk soil), to soils incubated without earthworms (Wolters and Joergensen 1992, Basker et al. 1992). It is not surprising that these various approaches produce different, and sometimes conflicting, results. Although we have attempted to make some generalizations about how casts and earthworm-worked soil differ from soils not ingested by earthworms, it is important to consider the design of the studies from which results are taken. Of particular importance are the choice of an appropriate reference material, and a time scale relevant to persistence of casts under field conditions (Martin and Marinissen 1993).

B.   Effects on Microbial Communities, Biomass, and Activity

Earthworm casts can be important soil microsites with high microbial abundance and activity, and often a different composition, compared to bulk soil. Barois and Lavelle (1986) found that *P. corethrurus* added large quantities of water and easily-assimilable organic matter to soils during transit through the gut, which resulted in large increases in microbial activity in the gut and smaller, but detectable, increases in fresh casts compared to noningested soil. Tiwari et al. (1989) reported higher microbial populations and enzyme activity (dehydrogenase, urease, and phosphatase) in field-collected earthworm casts than in bulk soil (sandy loam Oxisol). Dkhar and Mishra (1986) also reported greater populations of fungi, actinomycetes, and bacteria in field-collected casts than in the surrounding soil. Shaw and Pawluk (1986a) examined the fecal microbiology of an anecic species (*L. terrestris*) and two geophagous species (*Octolasion tyrtaeum* and *Aporrectodea turgida*) maintained in calcareous soils from three different textural classes (sandy loam, clay loam, and silty clay loam) incubated under controlled conditions for one year. At the end of the incubation, casts of *O. tyrtaeum* and *A. turgida* were grouped together and were analyzed separately from those of *L. terrestris*. Shaw and Pawluk (1986a) found that casts of both *L. terrestris* and *O. tyrtaeum/A. turgida* had densities of bacteria and actinomycetes which were 1-3 orders of magnitude higher than in control soils from incubation containers without earthworms. Densities of anaerobic microorganisms, filamentous fungi and *Cytophaga* also were higher in casts of both functional groups, although the magnitude of the differences between casts and control soil varied among soil types. Densities of yeasts were greater in casts of both functional groups in the clay loam soil, relative to the control soil, while in the sandy loam and silty clay loam soils only casts of *L. terrestris* exhibited increased densities of yeasts. The differential effects of soil texture on earthworm-induced changes in soil microbial communities under similar incubation conditions is noteworthy, and indicates that earthworm-microbial interactions can vary with soil type. Wolters and Joergensen (1992) also found that the effects of earthworms on soil microbial biomass and activity were highly dependent on soil type and quality of organic resources.

Earthworm-induced changes in microbial populations and activity vary with cast age. Scheu (1987) observed a transient increase in microbial biomass in freshly deposited casts of *Aporrectodea caliginosa*, while older casts supported less microbial biomass than did bulk soil (Figure 1A). Microbial biomass in casts was 128% of that in control soils within 4 h of being egested, and then decreased to approximately 90% of control soils after 2 weeks, and remained at 94% of control soils 30 days after egestion. However, microbial respiration remained elevated throughout the entire incubation, and was 86% higher, on average, in casts than in control soil. The simultaneous increase in respiration and decrease in microbial biomass suggests that casts of *A. caliginosa* may contain a smaller, but more metabolically active, microbial community than does bulk soil. These results are similar to those of Lavelle et al. (1992) for casts of *P. corethrurus* (Figure 1B). They noted a six to seven-fold increase in microbial biomass N in freshly egested casts, relative to noningested soil. However, within 12 h microbial biomass N in the casts had dropped to slightly more than twice the level in controls and then declined only slightly for the remainder of the 16 d incubation. The greater relative increase in microbial biomass observed by Lavelle et al. (1992), compared to Scheu (1987), may be explained in part by differences in organic matter content and microbial biomass of the bulk soils used. Scheu used a forest soil high in organic matter (7.6%) and microbial biomass (1162 μg C g^{-1}), while the soil used by Lavelle et al. was only 1.7% organic matter and supported a much lower microbial biomass (about 100 μg C g^{-1} based on a C:N ratio of 10). It seems likely that additions of labile C during gut passage can cause a transient increase in microbial biomass of fresh casts which is proportionally greater in soils naturally low in organic matter. This observation is supported by studies of the interactions of the tropical geophage *Millsonia anomala* and soluble soil organic matter. In these experiments Lavelle et al. (1980) manipulated the concentration of organic matter in soils fed to *M. anomala* by adding water-soluble organic matter (fresh grass extracts) to a low organic matter soil. Casts of earthworms feeding on soil with low concentrations of soluble organic matter had higher microbial activity than non-ingested soil, while casts of earthworms feeding on soils high in soluble organic matter exhibited decreased microbial activity.

Wolters and Joergensen (1992) also noted some interesting temporal shifts in the effects of earthworms (*A. caliginosa*) on soil microbial biomass and metabolic quotients following cessation of earthworm activity, although they did not isolate fresh casts from surrounding soil in these experiments. They incubated six different soils, with or without earthworms, for 21 days. At the

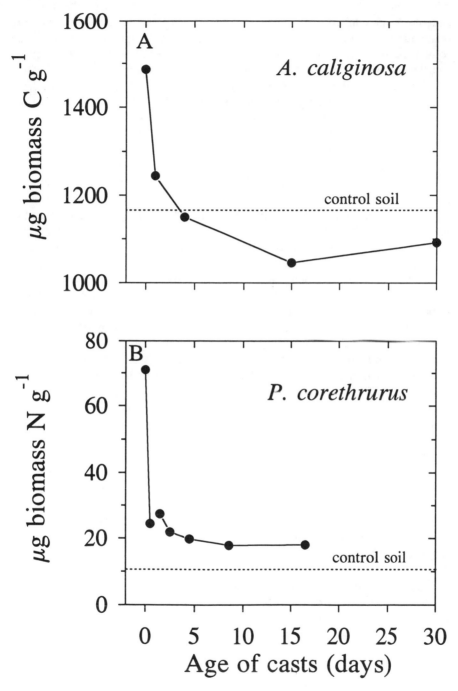

Figure 1. Changes in microbial biomass C over time in casts of *Aporrectodea caliginosa* (Figure 1A; from Scheu 1987) and changes in microbial biomass N over time in casts of *Pontoscolex corethrurus* (Figure 1B; from Lavelle et al. 1992).

end of 21 days microbial biomass and indices of specific metabolic activity (respiration per unit biomass) were measured in a subset of incubation containers, while in another subset of containers, the earthworms were removed and the earthworm-worked soil was incubated for an additional 21 days before measurements were taken. Immediately following the period of earthworm activity (21 days), microbial biomass was lower, relative to controls, in five of the six soils and microbial activity per unit biomass was increased in all soils. Changes in microbial biomass of earthworm-worked soils at 21 days were negatively correlated with initial organic matter content. This is consistent with the results of Lavelle et al. (1980), and with our comparisons of data from Scheu (1987) and Lavelle et al. (1992), providing further evidence that the effects of earthworms on microbial biomass and activity in casts in strongly influenced by organic matter content of the ingested soil. In contrast, Wolter and Joergensen (1992) found that the effects following cessation of earthworm activity and additional incubation (42 days) were highly soil specific, often were the opposite of those measured while earthworms were still active, and were not related to initial soil variables. Wolters and Joergensen (1992) speculate that the depression of microbial biomass and increase in specific microbial activity sometimes observed in fresh casts are due to an increase in the relative importance of bacteria and actinomycetes which can quickly mineralize labile organic substrates in fresh casts. As casts age, labile substrates are exhausted and fungi may become relatively more important, resulting in increased microbial immobilization of C and N, and a lower specific metabolic activity. The soil-specific, resource-specific, and time-dependent nature of microbial responses to earthworm activity may be responsible for some of the apparently conflicting results of various published studies of earthworm-microbial interactions.

One of the greatest impacts of macroinvertebrates, including earthworms, on microbial decomposers may be to alter fungal:bacterial ratios at a scale relevant to processes in the field (Anderson 1988). Potential changes, caused by earthworms, in relative proportions of soil fungi and bacteria are important because of differences in C assimilation efficiencies of fungi and bacteria (Adu and Oades 1978), and production of more recalcitrant organic substances by fungi which may affect soil organic matter (SOM) storage. However, most evidence for earthworm-induced changes in microbial community structure and activity comes from comparisons of fresh casts with control soils or from laboratory incubations of earthworms in small volumes of soil. The extent to which such effects occur in bulk soils in the field remains to be investigated, as does their functional significance.

C.   Interactions with Other Soil Invertebrates

It seems reasonable to conclude that earthworm-induced changes in the microbial community will affect higher trophic levels in the soil food web, and the

potential importance of earthworms in affecting populations of other invertebrates has been noted (Coleman 1985), but remains largely uninvestigated.
Earthworms may affect other soil-inhabiting invertebrates by altering their
resource base (amount and distribution of organic matter and microbial
populations), by affecting soil structure (porosity, aggregate structure), by direct
ingestion, and by dispersal. Alteration of the soil environment in microsites such
as earthworm middens can affect the distribution of soil invertebrates in some
ecosystems. Middens differ from surrounding soil in moisture content, water
holding capacity, nutrient status, and microbial composition and activity.
Hamilton and Sillman (1989), found differences in the microarthropod
communities associated with middens in fields and woodlots, and suggested that
middens contribute substantially to spatial variability in the distribution of soil
microarthropods. Marinissen and Bok (1988) reported that larger species of
Collembola, and larger individuals within a species, occurred in Dutch polders
that had been colonized by earthworms, compared to those without earthworms.
They attributed this to the presence of greater numbers of large pore spaces in
the presence of earthworms, although other factors such as distribution of
organic matter also may have been important. Earthworms are capable of
digesting ingested protozoa (Bamforth 1988), and may also affect nematode
populations by ingestion (Ellenby 1945, Dash et al. 1980) or by dispersion
(Shapiro et al. 1993). Shapiro et al. (1993) observed enhanced vertical dispersal
of the entomopathogenic nematode *Steinernema carpcapsae* in the presence of
L. terrestris and *Aporrectodea trapezoides*. They noted the presence of
nematodes on the exterior and interior of earthworms and suggested that
transport by earthworms may increase dispersal of entomopathogenic nematodes
in soil. Earthworms introduced into New Zealand pastures reduced total
nematode numbers by 37–66% (Yeates 1981). Although the generic composition
of the nematode community was unchanged, populations of bacterial feeders and
fungal-feeding Tylenchida were decreased in the presence of earthworms. This
may have been due to ingestion of free-living nematodes by earthworms or may
reflect changes in the soil microbial community caused by earthworms.
Unfortunately, there are few studies that have examined the effects of earthworms on other soil invertebrate groups and on populations and activity of soil
microorganisms concurrently. Earthworm-induced changes in the decomposer
community may affect decomposition and nutrient cycling processes and are in
need of further investigation.

D. Influences on Nutrient Transformations and Availability

Availability of nutrients (C, N, P, K, Ca) in earthworm casts and burrow walls
is generally higher than in bulk soils, and many studies have indicated that
earthworm casts and burrows are important microsites for some specific nutrient
transformations. The quantity, form, and distribution of C in earthworm casts
and burrow linings is often quite different than in bulk soil. Lee (1985) reviewed

data from a number of studies comparing casts to bulk soils and found consistently higher quantities of C in casts. In a heterogenous soil environment, some of this effect can be attributed to selective feeding and preferential ingestion of soil higher in organic matter. However, it also is clear that earthworm feeding can modify the form and distribution of ingested organic matter. Barois and Lavelle (1986) examined some of the physicochemical changes that occur in the earthworm gut and noted marked changes in concentrations of water-soluble organic compounds during soil transit. They suggested that addition of readily-assimilable organic compounds during passage through the earthworm gut results in increased microbial degradation of more complex organic substrates in the soils. This "mutualistic digestive system" has been proposed for both tropical (Barois and Lavelle 1986, Barois 1992) and temperate (Trigo and Lavelle 1993) earthworm species. The form of carbon in casts also can be different than in bulk soil. Martin (1991) found that *M. anomala* reduced the coarse organic matter fraction (250–2000 μm) in fresh casts, relative to bulk soil, as a result of comminution of ingested organic matter. Shaw and Pawluk (1986a, b) reported greater amounts of clay-associated C in earthworm casts than in bulk soil, suggesting that earthworms may promote stabilization of soil C by promoting binding with clays.

Earthworms can significantly affect N transformations in casts and burrow walls. Levels of inorganic N are often quite high in fresh earthworm casts, with ammonium being the dominant form (Scheu 1987, Lavelle et al. 1992). However, nitrification usually occurs rapidly in casts, and several authors have noted a simultaneous decline in ammonium concentrations and increase in nitrate concentrations as casts age (Parle 1963, Syers et al. 1979). Lavelle et al. (1992), in studies with *P. corethrurus*, found that inorganic N concentrations in fresh casts were approximately five times greater than in non-ingested soil, with the greatest change occurring in ammonium concentrations (Figure 2). Concentrations of microbial biomass N in casts also were 6–7 times greater than in control soil. As the casts aged there was a rapid decline (approximately 50%) in concentrations of inorganic + microbial biomass N within the first 12 hours. For the remainder of the 16 d incubation, ammonium levels continued to decline, while nitrate levels increased. The increase in inorganic N concentration in casts, relative to non-ingested soils, is a combined result of excretion of ammonia into soil as it passes through the gut and increased microbial mineralization of organic N. The cause of the rapid decline in labile N in fresh casts is unknown, although ammonia volatilization and denitrification may be important and should be further investigated.

Relatively little is known regarding the effects of earthworms on gaseous N fluxes, although there is evidence that casts and burrow walls may be important microsites for both losses (denitrification, volatilization) and inputs (N-fixation). The combination of increased nitrate and organic C, soil aggregation, and microbial activity in casts can lead to conditions favorable for denitrification (Svensson et al. 1986). Elliott et al. (1990, 1991) compared denitrification rates from earthworm casts and bulk clay loam soil collected from fertilized and

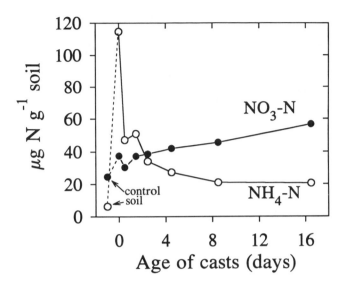

Figure 2. Temporal changes in concentrations of ammonium and nitrate nitrogen in aging casts of *Pontoscolex corethrurus* fed on an Amazonian Ultisol (from Lavelle et al. 1992).

unfertilized English pastures which were drained or undrained. They found consistently higher rates of denitrification in casts, compared to bulk soil samples, and concluded that earthworm casts are important microsites for denitrification and contribute to the spatial heterogeneity of denitrification measurements in the field. Knight et al. (1992) summarized the differences in N_2O production rates of casts and soils (Figure 3), and used these values together with data on annual rates of cast production to estimate annual denitrification rates from surface casts and bulk soils in these pastures. They estimated that surface casts could contribute from 12–26% of total denitrification annually.

In addition to the potential importance of casts as sites for gaseous loss of N, there also have been reports of earthworm activity increasing N-fixation rates. For example, Shaw and Pawluk (1986a) reported higher levels of acetylene reduction (an indication of N-fixation activity) by *Spirillum* sp. in burrow linings of *L. terrestris*, than in control soils. Bhatnagar (1975, reviewed in Lee 1985) reported that about 40% of the total aerobic N-fixers, 13% of the anaerobic N-fixers, and 16% of the denitrifying bacteria in the soil occurred within the 2 mm zone surrounding earthworm burrows (the drilosphere). Other studies also have suggested increased N fixation in the earthworm gut and/or casts (Spiers et al. 1986, Barois et al. 1987, Šimek and Pižl 1991). However, the extent to which earthworms enhance N fixation, and the potential significance of this phenomenon, remains debatable (Lee 1985) and needs further investigation.

Phosphorus availability in casts is often significantly greater than in bulk soils (Sharpley and Syers 1976, Tiwari et al. 1989, Krishnamoorthy 1990). In a U.S.

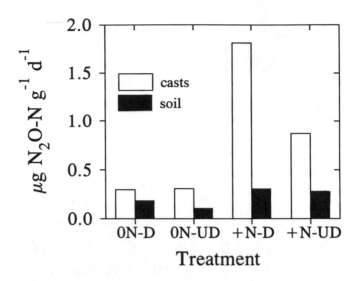

Figure 3. Mean rates of denitrification, expressed as nitrous oxide production, in field-collected casts and bulk soils of English pastures. Treatments are no fertilizer (0N) or 200 kg N ha^{-1} yr^{-1} (+N) and drained (D) or undrained (UD) (from Knight et al. 1992).

agroecosystem, Lunt and Jacobson (1944) reported that extractable P was 150 μg g^{-1} in casts, as opposed to 21 μg g^{-1} in bulk soil. James (1991) found that extractable P in surface casts of earthworms in a North American tallgrass prairie ecosystem usually ranged from 18–30 μg P g^{-1}, while bulk soils were in the range 3–10 μg P g^{-1}. Based on additions of C, N, and P to bulk soil, casts and burrow walls of *A. caliginosa*, Scheu (1987) concluded that P was less limiting to microbial activity in freshly deposited casts than in bulk soil or soil from burrow walls. However, P limitations tended to increase as casts aged, and this was reflected by a decrease in amounts of bicarbonate extractable P from 13.2 μg P g^{-1} in fresh casts to 7.6 μg P g^{-1} in 15-day-old casts. Increased P availability in fresh earthworm casts has been attributed to an increase in phosphatase activity in egested material (Satchell and Martin 1984), although it remains unclear to what extent the increased phosphatase activity is due directly to earthworm-derived enzymes, as opposed to increased microbial activity (Park et al. 1992). Other studies suggest that gut passage also alters the P-sorption capacities of some tropical soils (Lopez-Hernandez et al. 1993).

Earthworm activity may significantly affect calcium availability in soils. Many species of earthworms possess calciferous glands or esophageal regions which are involved in production of $CaCO_3$ spherules. Spiers et al. (1986) reported that a native North American earthworm in the genus *Arctiostrotus*, which occurs in the wet coastal forests of Vancouver Island, can convert calcium oxalate crystals on ingested fungal hyphae to calcium bicarbonate, which then is egested in cast material. This temporarily increases calcium availability in the fresh casts, and increases pH which could affect concentrations of other soluble nutrients available for plant uptake. Amounts of extractable potassium also have been reported to be elevated in earthworm casts, relative to bulk soils (Tiwari et al. 1989). In a laboratory experiment in which two individuals of *A. caliginosa* were maintained in 30 g of silt loam soil, Basker et al. (1992) found that earthworm activity increased the proportion of total soil potassium in exchangeable form, presumably by shifting the equilibrium between exchangeable and non-exchangeable forms in the soil.

E.   Effects of Earthworms in the Rhizosphere

The rhizosphere is another microsite where earthworms may have strong effects on nutrient cycling processes through their interactions with plant roots and rhizosphere microbes. Although there are few data available on earthworm-rhizosphere interactions, there appears to be potential for synergistic effects of root and earthworm activity on nutrient mineralization and plant uptake processes. Plant roots and earthworms affect soil structure, chemistry and microbial activity in similar ways. Both create new pores in the soil matrix, can increase soil aggregate stability, secrete labile low molecular weight C compounds, and can stimulate microbial activity. A key factor in earthworm-rhizosphere interactions could be the secretion of available or readily mineralizable N in earthworm mucus and casts. There have been few studies to determine if earthworm activity is greater in the rhizosphere than in bulk soil, but earthworm secretion of N in the rhizosphere would be available for root uptake with potential effects on plant production.

James and Seastedt (1986) present evidence for an earthworm-rhizosphere effect on N dynamics. They observed increased root biomass and lower levels of NO_3-N in leachate when *Diplocardia* spp. were present compared to pots without worms. They hypothesized that localized N mineralization in the rhizosphere may have increased plant uptake of N, contributing to both increased root growth and decreased N leaching. Spain et al. (1992) also provide evidence that earthworms increase plant production, and that the increased growth is associated with increased uptake of N and P in the presence of earthworms. They found positive correlations between earthworm biomass (*M. anomala* and two eudrilid species) and final plant biomass, and concentrations of N and P in plant roots. When microbial biomass was labelled with ^{15}N, earthworms increased the transfer of ^{15}N to both foliage and roots (Figure 4), suggesting that enhanced mineralization of N from microbial biomass in the rhizosphere could increase plant uptake. The ability of roots to utilize nutrients in earthworm casts has also been observed circumstantially by Spiers et al. (1986), who noted that live fine and very fine roots were concentrated in fresh casts of *Arctiostrotus* sp. in a Vancouver Island coastal forest.

Earthworms also may affect nutrient dynamics in the rhizosphere by consumption of live and/or dead roots. The importance of root consumption by earthworms rarely has been explored, but there is some evidence that both live and dead roots are ingested. James and Cunningham (1989) found root fragments among the gut contents of all seven species examined from a tallgrass prairie. With the exception of one species, the organic matter concentrations of the gut contents were nearly twice that of bulk soil, leading James and Cunningham (1989) to speculate that these worms were feeding preferentially in organically enriched sites and that well-decomposed material in the rhizosphere may be a preferred substrate. Whether or not earthworms consume live roots remains controversial, but there is some evidence this does occur. Cortez and Bouché (1992) used short-term labelling of rye-grass to enrich roots with ^{14}C. After 24 h of contact with plant roots, they found that gut contents of *L. terrestris* were labelled with ^{14}C and concluded that the worms had fed on root tissue. However, gut contents were not examined for root fragments and it was not possible to conclude definitively the existence of herbivory in earthworms. In a field experiment in which clover roots were labelled with ^{32}P applied to the foliage of the clover plants, Baylis et al. (1986) found high levels of ^{32}P in three species of earthworms after a six hour labelling period, suggesting that live root consumption occurred. The magnitude and potential significance of live-root feeding by earthworms will require further study.

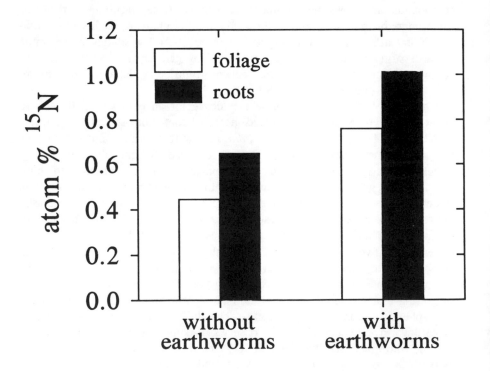

Figure 4. Effects of the earthworm *Milsonia anomala* on transfer of ^{15}N from soil microbial biomass to plants grown in an infertile savanna soil (from Spain et al. 1992).

VI.   Effects of Earthworms at the Ecosystem Level

The effects of earthworms on whole soil structure and function, and ecosystem-level nutrient fluxes, are less well understood than their effects at the microsite level. Some important research questions at this scale center on discerning the net effects of earthworms on soil structural development, storage and forms of soil organic matter and nutrients, influences of earthworm-induced macropores on fluxes of water and materials, and ecosystem-level estimates of nutrient inputs, retention, and losses.

Earthworms, through their burrowing and casting activities, turn over tremendous quantities of soil on an annual basis, and can induce major structural changes in soil which have important consequences for both water and nutrient fluxes. Soils with high populations of earthworms often have a greater number and altered size distribution of stable soil aggregates, greater porosity, higher infiltration rates, and less surface litter than soils without earthworms. Changes in soil aggregate structure and macropore densities and distribution may have significant effects on carbon and nutrient storage and movement in soils. In this section we focus on the effects of earthworms on soil organic matter dynamics, stable aggregate formation and its relationship to organic matter dynamics, the potential for earthworms to alter patterns and timing of nutrient availability, and the effects of earthworm-induced macropores on water and solute movement.

A.   Soil Organic Matter Dynamics

A fundamental question involving the effects of earthworms on soil organic matter is the extent to which they are able to utilize different organic matter fractions of the bulk soil. Earthworms ingest a range of organic matter fractions when feeding, but may differentially assimilate C from these various fractions. Martin et al. (1992a) used soils in which specific size classes were replaced with like size classes differentially labelled with ^{13}C, and found that M. anomala, a tropical geophage, assimilated soil organic matter from both large (250–2000 μm) size classes, considered to include newer plant-derived organic matter, and small (0–20 μm) size classes thought to comprise older clay-associated organic matter. However, other research utilizing $\delta^{13}C$ values of earthworms living in soils with differentially labelled organic pools has indicated that earthworms assimilate C primarily from young soil organic matter pools (Martin et al. 1992b). It appears that earthworms assimilate more readily decomposable organic substrates of the same age distribution utilized by the overall decomposer community. This seems to be true for both litter-feeding and geophagous species. Although geophages may ingest large quantities of older, humified organic matter, they apparently assimilate more C from the younger more labile fractions. These results are based, in part, on studies done in a temperate agroecosystem in France, and may apply to temperate ecosystems in North America, although this remains to be investigated.

In spite of their apparent propensity to utilize organic matter pools of similar age (Martin et al. 1992b), different species, or at least functional groups, of earthworms may have disparate effects on organic matter dynamics. Shaw and Pawluk (1986a) suggest that the effect of geophages in different soils is to enhance trends in decomposition which are already dictated by soil type, while anecic species, such as *L. terrestris*, may play a more prevalent role in modifying organic matter decomposition rates in different soils by removing surface litter and mixing it with mineral soil. A common perception is that earthworms, particularly anecic species, tend to increase organic matter decomposition rates. However, Shaw and Pawluk (1986a) found reduced decomposition rates of grass added to a sandy loam soil (lower soil respiration and greater increases in total soil C) in the presence of *L. terrestris*, relative to control soils or those where only geophages were present. The net effect was an increase in the fraction of C associated with the clay-bound fraction and a net increase in soil organic matter storage. In the same experiment, *L. terrestris* had the opposite effect on C dynamics in clay loam and silty clay loam soils, where decomposition of added grass residues was significantly accelerated.

One of the most basic unanswered questions regarding the effects of earthworms on soil organic matter dynamics is whether the net effect of earthworm activity is, in the long term, to increase rates of organic matter decomposition or to promote soil organic matter storage. There are numerous studies demonstrating accelerated breakdown of coarse particulate organic matter as a result of earthworm activity. For example, Mackay and Kladivko (1985) conducted a pot study in which soybean and maize residues were added to a Raub silt loam from Indiana with different densities of *L. rubellus*, and found that earthworms significantly decreased the amount of residue >300 μm recovered from the pots. Zachmann and Linden (1989) also observed lower recoveries of corn residues added to a Waukegan silt loam in Minnesota in the presence of *L. rubellus*. Parmelee et al. (1990), in a biocide field experiment in Georgia, found that reducing earthworm populations in conventional and no-tillage agroecosystems (Typic Rhodudult, Hiwassee sandy clay loam) resulted in increased standing stocks of both fine (790–2380 μm) and coarse (>2380 μm) particulate organic matter in the no-tillage system and coarse particulate organic matter in the conventional tillage system.

However, other recent research has suggested that stabilization of organic matter in earthworm casts may result in long-term reductions in mineralization of organic matter in some tropical soils, raising important questions about the long-term net effects of earthworms on soil organic matter pools. Martin (1991) examined the effects of the tropical geophagous earthworm *M. anomala* on soil organic matter dynamics over different time scales. In these experiments earthworms were maintained in homogenized and sieved soil. Fresh casts were compared to noningested soils to determine the short-term effects of soil ingestion on organic matter pools. Subsamples of collected casts then were placed in culture boxes and incubated for up to 420 days at constant temperature and moisture to determine the longer-term effects of feeding and cast production

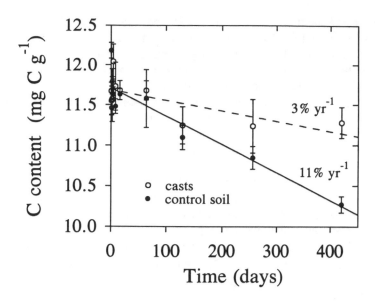

Figure 5. Changes in the C content of casts of *Milsonia anomala* and of unaggregated control soil (2 mm sieved) during 420 days of incubation at 15% moisture and 28°C (from Martin 1992).

on retention of carbon. Short-term effects of earthworm feeding included a significant reduction of organic matter contained in particle-size fractions between 250 and 2000 μm, and a 2% decline in total C content, suggesting that, in the short term, *M. anomala* increases comminution and mineralization of organic matter. However, the longer-term incubations of casts and control soils (Figure 5) revealed much lower C mineralization from the casts (3% yr^{-1}), than from the nonaggregated control soils (11% yr^{-1}). The reduction in C mineralization was attributed to increased protection of organic matter in stable soil aggregates created by earthworm casting. This may be an important mechanism for organic matter stabilization in tropical soils where abiotic factors (climate and soil texture) favor rapid C mineralization (Lavelle and Martin 1992).

The extent to which earthworms can increase protection of organic matter in temperate soils is largely unknown. There are conflicting data, including some from North America, regarding the effects of earthworms on formation and persistence of stable soil aggregates in temperate soils. Mackay and Kladivko (1985) reported that *L. rubellus* significantly increased the water stability of moist soil aggregates in a Raub silt loam soil incubated with or without crop residues. However, this effect was not apparent when soils were air-dried and rewetted. Shipitalo and Protz (1988) found that clay dispersability of fresh casts was greater than that of bulk soil, but aging increased stability of casts. They also provide data suggesting that the quality of organic matter ingested is a determinant of cast stability. West et al. (1991) showed that *L. rubellus* increased the proportion of water-stable aggregates in the size fraction >2 mm and decreased the proportion between 0.25 and 2 mm in laboratory incubations with two Ultisols from the Georgia Piedmont. The effect was greater in a Cecil loamy sand than in a Pacolet sandy clay loam. Preliminary data from an experiment in Ohio in which earthworm populations are being manipulated (reduced, enhanced, unmanipulated) in field enclosures (Blair et al. in press) suggests that earthworms can affect the stability of soil aggregates under field conditions (Ketterings 1992). Data from the first year of the experiment show that experimental increases in earthworm populations resulted in increased water-stability of aggregates in the three largest size classes examined (1–2, 2–4, and 4–10 mm). Earthworm reductions decreased the proportion of water-stable aggregates in the largest size class only. The increase in the proportion of large water-stable aggregates under enhanced earthworm populations was associated with increased C and N content of the aggregates. Further long-term studies are needed to determine the extent to which earthworms influence the production and distribution of water-stable soil aggregates and the impact this has on soil organic matter dynamics in North American soils. Two important considerations are the form of the casts (compact or slurries) and the time scales over which they persist in the field.

B.   Effects on Plant Productivity

Net soil organic matter accumulation or degradation depends on the balance between production and decomposition. Most of our discussion so far has focused on the potential of earthworms to alter soil organic matter dynamics by influencing decomposition and mineralization processes (comminution, redistribution, aggregate formation, physical protection). However, the potential of earthworms to affect plant production, and rates of organic inputs, also should be considered. There are data indicating that earthworms can increase plant productivity in pot experiments (i.e., Hopp and Slater 1949, Spain et al. 1992), although this effect has not been observed in all experiments (i.e., Mackay and Kladivko 1985, James and Seastedt 1986). There are fewer data regarding the effects of earthworms on plant productivity under field conditions, although a stimulatory effect has been noted in earthworm introduction experiments in New Zealand pastures (Stockdill 1982) and reclaimed Dutch polders (Hoogerkamp et al. 1983). Earthworms may influence plant growth by altering physical characteristics of the soil and soil nutrient availability. In addition, several studies have pointed to the presence of growth regulating substances in earthworm casts as a possible mechanism affecting plant growth (Nielson 1965, Atlavinyte and Dačiulyte 1969, Springett and Syers 1979, Krishnamoorthy and Vajranabhaiah 1986), although the specific effects of these substances, and their relationships with plant productivity in the field, are still not well understood (Tomati et al. 1988).

C.   Spatial and Temporal Patterns of Nutrient Availability

In earlier sections we considered the effects of earthworms on nutrient availability in specific microsites (i.e., casts). Relating these microsite effects to measurable changes in nutrient availability in bulk soils and ecosystem-level nutrient cycling processes is a formidable challenge. Earthworm activities appear to have the potential to alter patterns and timing of nutrient availability, particularly in temperate ecosystems where earthworm activity exhibits a definite seasonality. We suggest that an important effect of earthworms may be to alter patterns of microbial mineralization-immobilization, although this does not appear to have been examined in detail. In a study of N dynamics in earthworm casts in a New Zealand pasture, Syers et al. (1979) found that the production of surface casts (predominantly from *L. rubellus*,) and concentrations of C, total N, and inorganic N in the casts, followed clear seasonal patterns. Peak total N and C content of casts coincided with peak litter production. Concentrations of ammonium in the casts at the beginning of the casting season (April) were approximately 80 μg N g^{-1}, but dropped to 55 μg N g^{-1} by mid May. Concentrations of ammonium then increased to a peak of about 115 μg N g^{-1} in August, before declining to a level similar to that at the start of the study. Changes in nitrate concentration were inversely related to ammonium during this period and increased from about 25 to 48 μg N g^{-1}. These patterns appear to be more closely related to cast production and litter removal, than to seasonal variation in soil nutrient pools. Although the accumulation of inorganic N in casts is small (3.5 kg N ha^{-1} yr^{-1}) compared to total system turnover of N, timing and spatial distribution of this N may magnify its importance. James (1991), found that the N and P content of surface casts of earthworms in North American tallgrass prairie ecosystems also varied seasonally. Ammonium and extractable P concentrations were higher in the spring and fall than in midsummer (Figure 6). Such combinations of higher casting rates and greater concentrations of extractable N and P in the spring and fall could alter patterns of N and P availability, although the magnitude of this effect and its potential importance for nutrient cycling and retention are unknown.

In addition to differences in seasonal cast production, Wolters and Joergensen (1992) suggest another mechanism by which earthworms could influence patterns of N mineralization and immobilization. They found that ongoing earthworm activity reduced soil microbial biomass, and therefore microbial immobilization of N. However, as earthworm-worked soil aged following the cessation of earthworm activity, N immobilization in microbial biomass increased in at least some soil types, suggesting the possibility of enhanced N mineralization coinciding with seasonal periods of earthworm activity and increased immobilization as earthworms become inactive. We suggest that further studies of temporal shifts in nutrient availability in response to earthworm activity, in a variety of terrestrial ecosystems, are essential for understanding the influence of earthworms on biogeochemical processes at the ecosystem level.

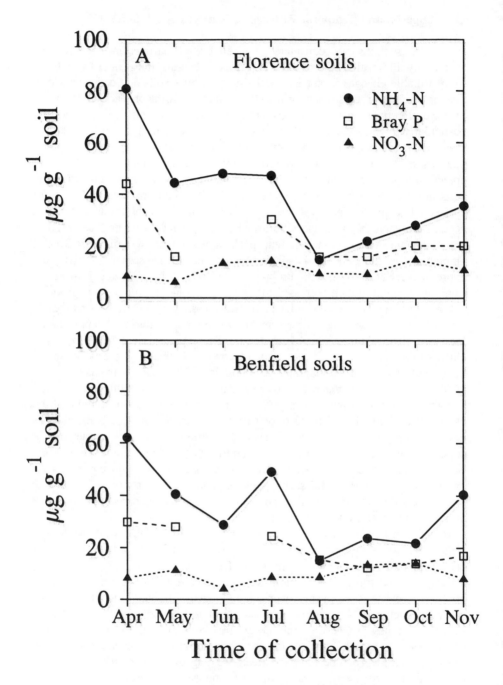

Figure 6. Mean monthly concentrations of NH_4-N, NO_3-N and Bray P in surface collected casts from Florence (4A) and Benfield (4B) soils at a Kansas tallgrass prairie site (from James 1991).

D.   Water and Solute Transport

A final aspect to consider is the effect of earthworms on water and solute movement in soils. Earthworm burrows may create preferential flow pathways in the soil profile which alter water balance and water movement (See Tomlin et al. (Chapter 7) this volume). These changes can have important consequences for movement of solutes in soils, and leaching of nutrients and pesticides into subsurface soil and ground water below agroecosystems. Zachmann and Linden (1989) found that *L. rubellus* could significantly increase the depth to which surface applied bromide moved, although the effect was dependent on the presence of surface applied corn residues. They concluded that *L. rubellus* forms stable burrows with many surface openings when surface residues are present, and that these burrows can decrease surface runoff and increase infiltration. Anecic species, which form well-developed vertical burrows up to several meters deep, may be particularly important in creating macropores which can channel water and solutes deep into the soil profile (Edwards et al. 1989, 1992). By increasing infiltration rates earthworms may reduce overland flow and associated losses of soil and nutrients. In one of the few studies investigating the effects of earthworms on overland transport of nutrients, Sharpley et al. (1979) observed increased losses of N and P in surface runoff in New Zealand pastures treated with carbaryl to eliminate earthworms, compared to untreated pastures where earthworms were present. Removal of earthworms resulted in a two-fold increase in volume of surface runoff, and even higher proportional increases in loss of dissolved inorganic N and P. In contrast, the transport of particulate P in surface runoff was greater in pastures where earthworms were present, and was associated with increased sediment transport in the presence of surface casting earthworms. Thus, an assessment of the net effect of earthworms on hydrologic export of nutrients should include effects on both surface and subsurface flows.

VII. Summary

It is surprising how little is actually known about the net effects of earthworms on the biogeochemistry of terrestrial ecosystems in North America. As we have illustrated, there have been some excellent studies on the roles of earthworms in various aspects of nutrient cycling processes in North America and elsewhere. Certainly because of ease, convenience, and practicality, many earthworm researchers have concentrated on earthworm effects at the microsite level with studies conducted over relatively short time scales, often in artificial laboratory microcosms. In our discussion of the effects of earthworms on microbial abundance, composition, activity, and nutrient transformations in casts, we have attempted to draw some general conclusions, such as the tendency for nutrient availability and microbial activity to be enhanced in casts, compared to bulk soil. However, it is possible to find contradictory evidence for almost every parameter examined. This is not for lack of good experimental data, but rather because results are dependent on the species of earthworm studied, as well as a host of other variables (soil type, physicochemical variables, organic matter availability, etc.), and the spatiotemporal scale at which measurements were made.

While earthworms clearly influence transformations of organic matter and nutrients in the soil, the relationships of these effects to ecosystem-level nutrient fluxes, and long-term accumulation or loss of nutrients, are not well understood. It appears that earthworms contribute to nutrient cycling processes in ways that can lead to either increased loss or increased storage of nutrients. Considering carbon, for example, the effect of earthworms usually is to accelerate the breakdown of coarse organic matter and contribute directly and indirectly to increased soil respiration, processes leading to loss of C from the system. But evidence also exists that egestion of C in casts can lead to increased storage through protection of organic matter in stable aggregates and by binding with clays. With regard to nitrogen, earthworms may increase N uptake by microbes and plants, alter the timing and location of N availability, and perhaps even increase N inputs to the system by fixation. However, there also is evidence that earthworm activity contributes to system loss of N through denitrification, volatilization, erosion, and leaching. Ultimately, the most important question regarding the influence of earthworms on biogeochemistry and ecosystem structure and function may be how they influence the balance between processes leading to loss and storage of C and other nutrients.

VIII.   Research Imperatives

Spatially and temporally explicit studies of the effects of earthworms on biogeochemical processes are needed at a variety of scales, from the earthworm gut to ecosystem-level effects, and in a variety of ecosystem types. Several general questions to be addressed include: How do earthworms alter the spatial and temporal scales at which nutrient cycling processes occur? What is the relative importance of short-term effects (i.e., rapid recycling of N) and longer-term effects (i.e., storage of C and N) of earthworm activity? How important are microsite effects to overall ecosystem function? A difficult challenge will be the integration of studies across a range of scales to adequately address both mechanistic explanations of the effects of earthworms on nutrient transformations and ecosystem-level consequences of earthworm activity for biogeochemical cycling. Some specific areas in which further research is needed are listed below.

A.   Gut Physiology and Biochemistry

Further research is needed on the physiological and biochemical transformations taking place in the earthworm gut, especially as they relate to changes in nutrient availability and microbial processes in casts. Specific gaps include information on C and N assimilation efficiencies of earthworm species with different feeding preferences, and further data on primary sources of C and nutrients for earthworms (i.e., bacteria, fungi, specific organic fractions). Also, further research focusing on the calcium relationships of earthworms, and the relationship of calcium release to availability of other nutrients should be explored.

B.   Functional Groups

Further studies of the species-specific effects of earthworms at different scales are needed, and could lead to a redefinition of functional groups in earthworms. There is already considerable evidence that different earthworm species have dissimilar effects on soil structure and nutrient transformations. This raises some important questions about the functional similarities of earthworms grouped into broad ecological categories. Are there better ways of defining ecological categories of earthworms? Do the same functional groupings apply to both temperate and tropical ecosystems?

C.   Native Species

More studies of the effects of native species, and comparisons of the roles of native and introduced species where both occur, are needed. Even the basic ecology of most native North American earthworms is poorly known. James (1991) has demonstrated that in tallgrass prairie ecosystems, native North American earthworms can affect nutrient cycling processes differently than introduced European species. Because the displacement of native species by exotics is a relatively recent, and sometimes ongoing, phenomenon (see Kalisz and Wood, Chapter 5, this volume), this should be explored in other ecosystems types.

D.   Isotope Studies

Expanded use of stable and radioactive isotopes are needed in both field and laboratory experiments. Use of enriched materials (^{13}C, ^{14}C, ^{32}P, and ^{15}N) has been an effective tool for determining the effects on earthworms on the fate on organic substrates and pathways of nutrient transformation (Cortez et al. 1989, Spain et al. 1992, Binet and Trehen 1992). Analysis of $\delta^{13}C$ values of earthworm tissue has provided some unique insights into the utilization of specific organic matter pools by earthworms in temperate European and tropical ecosystems (Martin et al. 1992a, b). There appear to be opportunities for similar studies in many North American ecosystems, which could provide further insights into the relationships of earthworms with specific organic matter fractions.

E.   Nutrient Availability

Studies of the relationship between earthworm activity and spatial and temporal patterns of nutrient availability are needed, particularly in temperate North American ecosystems. We have pointed out several ways in which earthworms influence rates of nutrient transformation and patterns of nutrient availability at the microsite level. The net effect of these microsites on spatiotemporal patterns of nutrient availability in the field and potential for nutrient loss from ecosystems remains largely uninvestigated. Two specific areas requiring further research are the effects of earthworm activity on gaseous N fluxes (N fixation, volatilization, and denitrification) and the role of earthworm casting in aggregate formation and persistence in relationship to storage of C and N. Additionally, research directed at the effects of earthworms on elements other than C and N is needed.

F.   Plant Growth

Quantitative studies of the effects of earthworms on plant growth and productivity under field conditions are needed. Earthworm-microbial-rhizosphere interactions seem to have particular promise for future studies. Indications that earthworm activity can be concentrated in the rhizosphere and the potential significance of root herbivory by earthworms should be examined.

G.   Interactions with Soil Biota

Earthworms have the potential to affect other components of the soil community in many ways, thereby altering both the structure and function of decomposer food webs and, presumably, nutrient transformations mediated by the decomposer community. To date there have been few studies which have examined simultaneously the effects of earthworms on microbial and invertebrate populations and activity. Further research into the effects of earthworms on soil microbial and invertebrate communities is recommended.

H.   Spatial Scales

Appropriate techniques for scaling up from effects observed at the microsite (i.e., casts and burrow linings) to effects on whole soil structure and function are needed. Extrapolation of results from microsites or laboratory incubations to field-scale estimates of nutrient flux will require innovative approaches. Quantification of the densities of specific microsites, and rates of cast, burrow and midden production and disappearance, in different ecosystem types may be required. In addition, landscape-level approaches which explicitly consider the interactions of earthworms with factors such as soil type, texture and resource quality may be particularly useful in scaling from the microsite to the ecosystem or landscape level.

I.   Long Term Studies

Finally, there is a definite need for further long-term manipulative field experiments (eliminations, additions) to assess the net effects of earthworms on SOM and nutrient dynamics in a variety of ecosystem types. As indicated in this chapter and in other reviews, there are considerable data on the shorter-term effects of earthworms at the microsite level, and on soils and soil-plant relationships in laboratory and greenhouse incubations. There is much less information on the extent to which these effects occur in the field, their magnitude and their importance to ecosystem-level nutrient fluxes. We suggest that these questions will be adequately addressed only through the use of field experiments in which entire earthworm communities are manipulated for an extended time. We are aware of such experiments currently being done in Ohio, Georgia, and La Mancha, Mexico, in North America. All of these studies are being done in agricultural ecosystems, and there is a need for similar studies to be initiated in other ecosystem types. In addition to manipulating pre-existing earthworm communities, it may be possible to find areas in North America where reintroduction of earthworms following glaciation is ongoing (see James, Chapter 2, this volume). In these areas, studies of long-term changes following reintroduction of earthworms could provide new insights into the net effects of earthworms on a variety of ecosystem properties and processes.

Acknowledgments

We thank G.G. Brown for his input during revision of the manuscript. Support for preparation of this chapter was provided, in part, by NSF grant DEB-9020461. Contribution no. 94-429-B from the Kansas Agricultural Experiment Station.

References

Adu, J.K. and J.M. Oades. 1978. Utilization of organic materials in soil aggregates by bacteria and fungi. *Soil Biol. Biochem.* 10:117–122.

Anderson, J.M. 1988. Spatiotemporal effects of invertebrates on soil processes. *Biol. Fertil. Soils* 6:216–227.

Atlavinyte, O. and J. Dačiulyte. 1969. The effect of earthworms on the accumulation of vitamin B_{12} in soil. *Pedobiologia* 9:165–170.

Bamforth, S.S. 1988. Interactions between protozoa and other organisms. *Agric. Ecosys. Environ.* 24:229–234.

Barois, I. 1992. Mucus production and microbial activity in the gut of two species of *Amynthas* (Megascolecidae) from cold and warm tropical climates. *Soil Biol. Biochem.* 24:1507–1510.

Barois, I. and P. Lavelle. 1986. Changes in respiration rate and some physicochemical properties of a tropical soil during transit through *Pontoscolex corethrurus* (Glossoscolecidae, Oligochaeta). *Soil Biol. Biochem.* 18:539–541.

Barois, I., B. Verdier, P. Kaiser, A. Mariotti, P. Rangel, and P. Lavelle. 1987. Influence of the tropical earthworm *Pontoscolex corethrurus* (Glossoscolecidae) on the fixation and mineralisation of nitrogen. p. 151–158. In: Bonvicini Pagliai, M. and P. Omodeo (eds.) *On earthworms*. A. Mucchi, Bologna, Italy.

Basker, A., A.N. Macgregor, and J.H. Kirkman. 1992. Influence of soil ingestion by earthworms on the availability of potassium in soil: An incubation experiment. *Biol. Fertil. Soils* 14:300–303.

Baylis, J.P., J.M. Cherrett, and J.B. Ford. 1986. A survey of the invertebrates feeding on living clover roots (*Trifolium repens* L.) using ^{32}P as a radiotracer. *Pedobiologia* 29:201–208.

Bhatnagar, T. 1975. Lombricienes et humification: Un aspect nouveau de l'incorporation microbienne d'azote induite par les vers de terre. p. 169–182. In: Gilbertus, K., O. Reisinger, A. Mourey, and J.A. Cancela de Fonseca (eds.) *Biodégradation et humification*. Pierron, Sarreguemines.

Binet, F. and P. Trehen. 1992. Experimental microcosm study of the role of *Lumbricus terrestris* (Oligochaeta: Lumbricidae) on nitrogen dynamics in cultivated soils. *Soil Biol. Biochem.* 24:1501–1506.

Blair, J.M., P.J. Bohlen, C.A. Edwards, B.R. Stinner, D.A. McCartney, and M.F. Allen. (in press) Manipulation of earthworm populations in field experiments in agroecosystems. *Acta Zoologica Fennica*.

Bolton, P.J. and J. Phillipson. 1976. Burrowing, feeding, egestion and energy budget of *Allolobophora rosea* (Savigny) (Lumbricidae). *Oecologia (Berlin)* 23:225–245.

Bouché, M.B. 1975. Action de la faune sur les états de la matierère organique dans les écosystèmes. p. 157–168. In: Gilbertus, K., O. Reisinger, A. Mourey, and J.A. Cancela da Fonseca (eds.) *Biodégradation et humification*. Pierron, Sarruguemines.

Bouché, M.B. 1977. Stratégies lombriciennes. In: Lohm, U. and T. Persson (eds.) *Soil organisms as components of ecosystems*. Ecological Bulletin (Stockholm) 25:122–132.

Christensen, O. 1988. The direct effects of earthworms on nitrogen turnover in cultivated soils. *Ecological Bulletin* 39:41–44.

Coleman, D.C. 1985. Through a ped darkly: an ecological assessment of root-soil-microbial-faunal interactions. p. 1–21. In: Fitter, A.H., D. Atkinson, D.J. Read, and

M.B. Usher (eds.) *Ecological interactions in soil*. Blackwell Scientific Publishing, London.

Cortez, J. and M.B. Bouché. 1992. Do earthworms eat living roots? *Soil Biol. Biochem.* 24:913–915.

Cortez, J., R. Hamheed, and M.B. Bouché. 1989. C and N transfer in soil with or without earthworms fed with ^{14}C- and ^{15}N-labelled wheat straw. *Soil Biol. Biochem.* 21:491–497.

Crossley, D. A. Jr., R.E. Reichle, and C.A. Edwards. 1971. Intake and turnover of cesium by earthworms (Lumbricidae). *Pedobiologia* 11:71–76.

Darwin, C. 1881. *The formation of vegetable mould through the action of worms with observations on their habits*. Faber and Faber, London.

Dash, M.C., B.K. Senapati, and P.C. Mishra. 1980. Nematode feeding by tropical earthworms. *Oikos* 34:322–325.

Dickschen, F. and W. Topp. 1987. Feeding activities and assimilation efficiencies of *Lumbricus rubellus* (Lumbricidae) on a plant only diet. *Pedobiologia* 30:31–37.

Dkhar, M.S. and R.R. Mishra. 1986. Microflora in earthworm casts. *J. Soil Biol. Ecol.* 6:24–31.

Edwards, C.A. and J.R. Lofty. 1977. *Biology of earthworms*. 2nd ed. Chapman and Hall, London.

Edwards, W.M., M.J. Shipitalo, L.B. Owens, and L.B. Norton. 1989. Water and nitrate movement in earthworm burrows within long-term no-till corn fields. *J. Soil Water Cons.* 44:240–243.

Edwards, W.M., M.J. Shipitalo, S.J. Traina, C.A. Edwards, and L.B. Owens. 1992. Role of *Lumbricus terrestris* (L.) burrows on quality of infiltrating water. *Soil Biol. Biochem.* 24:1683–1689.

Ellenby, C. 1945. Influence of earthworms on larval emergence in the potato root eelworm *Heterodera rostochiensis* Wollenweber. *Ann. Appl. Biol.* 31:332–339.

Elliott, P.W., D. Knight, and J.M. Anderson. 1990. Denitrification in earthworm casts and soil from pasture under different fertilizer and drainage regimes. *Soil Biol. Biochem.* 22:601–605.

Elliott, P.W., D. Knight, and J.M. Anderson. 1991. Variables controlling denitrification from earthworm casts and soil in permanent pasture. *Biol. Fertil. Soils* 11:24–29.

Ferriere, G. and M.B. Bouché. 1985. Première mesure écophysiologique d'un débit d'azote de *Nicodrilus longus* (Ude) (*Lumbricidae* Oligochaeta) dans la prairie de Citeaux. *CR Academy Science* 301:789–794.

Hamilton, W.E. and D.Y. Sillman. 1989. Influence of earthworm middens on the distribution of soil microarthropods. *Biol. Fertil. Soils* 8:279–284.

Hendrix, P.F., D.A. Crossley Jr., D.C. Coleman, R.W. Parmelee, and M.H. Beare. 1987. Carbon dynamics in soil microbes and fauna in conventional and no-tillage agroecosystems. *INTECOL Bulletin* 15:59–63.

Hoogerkamp, M., H. Rogaar, and H.J.P. Eijsackers. 1983. Effect of earthworms on grassland on recently reclaimed polder soils in the Netherlands. p. 85–105. In: Satchell, J.E. (ed.) *Earthworm ecology. From Darwin to vermiculture*. Chapman and Hall, New York.

Hopp, H. and C.S. Slater. 1949. The effect of earthworms on the productivity of agricultural soil. *J. Agric. Res.* 78:325–339.

James, S.W. 1991. Soil, nitrogen, phosphorus, and organic matter processing by earthworms in tallgrass prairie. *Ecology* 72:2101–2109.

James, S.W. and M.R. Cunningham. 1989. Feeding ecology of some earthworms in Kansas tallgrass prairie. *Am. Midl. Nat.* 121:78–83.

James, S.W. and T.R. Seastedt. 1986. Nitrogen mineralization by native and introduced earthworms: Effects on big bluestem growth. *Ecology* 67:1094–1097.

Judas, M. 1992. Gut content analysis of earthworms (Lumbricidae) in a beechwood. *Soil Biol. Biochem.* 24:1413–1417.

Ketterings, Q.M. 1992. Effects of earthworm activities on soil structure stability and soil carbon and nitrogen storage in organic-based and conventional agroecosystems. M.Sc. thesis, Wageningen Agricultural University, The Netherlands. 136 pp.

Knight, D., P.W. Elliott, J.M. Anderson, and D. Scholefield. 1992. The role of earthworms in managed, permanent pastures in Devon, England. *Soil Biol. Biochem.* 24:1511–1518.

Krishnamoorthy, R.V. 1990. Mineralization of phosphorus by faecal phosphatases of some earthworms of Indian tropics. *Proc. Indian Acad. Sci. (Animal Sci.)* 99:509–518.

Krishnamoorthy, R.V. and S.N. Vajranabhaiah. 1986. Biological activity of earthworm casts: an assessment of plant growth promoter levels in the casts. *Proc. Indian Acad. Sci. (Animal Sci.)* 95:341–351.

Lavelle, P. 1988. Earthworm activities and the soil system. *Biol. Fertil. Soils* 6:237–251.

Lavelle, P. and A. Martin. 1992. Small-scale and large-scale effects of endogeic earthworms on soil organic matter dynamics in soils of the humid tropics. *Soil Biol. Biochem.* 24:1491–1498.

Lavelle, P., G. Melendez, B. Pashanasi, and R. Schaefer. 1992. Nitrogen mineralization and reorganization in casts of the geophagous tropical earthworm *Pontoscolex corethrurus* (Glossoscolecidae). *Biol. Fertil. Soils* 14:49–53.

Lavelle, P., B. Sow, and R. Schaefer. 1980. The geophagous earthworm community in the Lamto savanna (Ivory Coast): niche partitioning and utilization of soil nutritive resources. p. 653–672. In: Dindal, D. (ed.) *Soil biology as related to land use practices.* EPA, Washington, D.C.

Lee, K.E. 1985. *Earthworms: their ecology and relationships with soils and land use.* Academic Press, New York.

Lopez-Hernandez, D., P. Lavelle, J.C. Fardeau, and M. Niño. 1993. Phosphorus transformations in two-P-sorption contrasting tropical soils during transit through *Pontoscolex corethrurus* (Glossoscolecidae: Oligochaeta). *Soil Biol. Biochem.* 25:789–792.

Lunt, H.A. and G.M. Jacobson. 1944. The chemical composition of earthworm casts. *Soil Sci.* 58:367–375.

Mackay, A.D. and E.J. Kladivko. 1985. Earthworms and rate of breakdown of soybean and maize residues in soil. *Soil Biol. Biochem.* 17:851–857.

Marinissen, J.C.Y. and J. Bok. 1988. Earthworm-amended soil structure: Its influence on Collembola populations in grassland. *Pedobiologia* 32:243–252.

Martin, A. 1991. Short- and long-term effects of the endogeic earthworm *Millsonia anomala* (Omodeo) (Megascolicidae, Oligochaeta) of tropical savannas, on soil organic matter. *Biol. Fertil. Soils* 11:234–238.

Martin, A., J. Balesdent, and A. Mariotti. 1992b. Earthworm diet related to soil organic matter dynamics through ^{13}C measurement. *Oecologia* 91:23–29.

Martin, A. and J.C.Y. Marinissen. 1993. Biological and physico-chemical processes in excrements of soil animals. *Geoderma* 56:331–347.

Martin, A., A. Mariotti, J. Balesdent, and P. Lavelle. 1992a. Soil organic matter assimilation by a geophagous tropical earthworm based on $\delta^{13}C$ measurements. *Ecology* 73:118–128.

Nielson, R.L. 1965. Presence of plant growth substances in earthworms demonstrated by paper chromatography and the Went pea test. *Nature* 208:113–114.

Park, S.C., T.J. Smith, and M.S. Bisesi. 1992. Activities of phosphomonoesterase and phosphodiesterase from *Lumbricus terrestris*. *Soil Biol. Biochem.* 24:873–876.

Parle, J.N. 1963. Microorganisms in the intestines of earthworms. *J. Gen. Microbiol.* 31:1–11.

Parmelee, R. W., M.H. Beare, W. Cheng, P.F. Hendrix, S.J. Rider, D.A. Crossley Jr., and D.C. Coleman. 1990. Earthworms and enchytraeids in conventional and no-tillage agroecosystems: A biocide approach to assess their role in organic matter breakdown. *Biol. Fertil. Soils* 10:1–10.

Parmelee, R.W. and D.A. Crossley, Jr. 1988. Earthworm production and role in the nitrogen cycle of a no-tillage agroecosystem on the Georgia piedmont. *Pedobiologia* 32:351–361.

Satchell, J. E. 1963. Nitrogen turnover by a woodland population of *Lumbricus terrestris*. p. 60–66. In: Doeksen, J. and J. van der Drift (eds.) *Soil organisms*. North-Holland Publishing Co., Amsterdam.

Satchell, J.E. 1967. Lumbricidae. p. 259–322. In: Burges, A. and F. Raw (eds.) *Soil biology*. Academic Press, London.

Satchell, J.E. (ed.) 1983. *Earthworm ecology. From Darwin to vermiculture*. Chapman and Hall, New York.

Satchell, J.E. and K. Martin. 1984. Phosphatase activity in earthworm faeces. *Soil Biol. Biochem.* 16:191–194.

Scheu, S. 1987. Microbial activity and nutrient dynamics in earthworm casts (Lumbricidae). *Biol. Fertil. Soils* 5:230–234.

Shapiro, D.I., E.C. Berry, and L.C. Lewis. 1993. Interactions between nematodes and earthworms: Enhanced dispersal of *Steinernema carpocapsae*. *J. Nematology* 25:189–192.

Sharpley, A.N. and J.K. Syers. 1976. Potential role of earthworm casts for the phosphorus enrichment of runoff waters. *Soil Biol. Biochem.* 8:341–346.

Sharpley, A.N., J.K. Syers, and J.A. Springett. 1979. Effect of surface-casting earthworms on the transport of phosphorus and nitrogen in surface runoff from pasture. *Soil Biol. Biochem.* 11:459–462.

Shaw, C. and S. Pawluk. 1986a. Faecal microbiology of *Octolasion tyrtaeum*, *Aporrectodea turgida* and *Lumbricus terrestris* and its relation to carbon budgets of three artificial soils. *Pedobiologia* 29:377–389.

Shaw, C. and S. Pawluk. 1986b. The development of soil structure by *Octolasion tyrtaeum, Aporrectodea turgida* and *Lumbricus terrestris* in parent materials belonging to different textural classes. *Pedobiologia* 29:327–339.

Shipitalo, M.J. and R. Protz. 1988. Factors influencing the dispersability of clay in worm casts. *Soil Sci. Soc. Am. J.* 52:764–769.

Šimek, M. and V. Pižl. 1989. The effects of earthworms (Lumbricidae) on nitrogenase activity in soil. *Biol. Fertil. Soils* 7:370–373.

Spain, A. V., P. Lavelle, and A. Mariotti. 1992. Stimulation of plant growth by tropical earthworms. *Soil Biol. Biochem.* 24:1629–1633.

Spiers, G.A., D. Gagnon, G.E. Nason, E.C. Packee, and J.D. Lousier. 1986. Effects and importance of indigenous earthworms on decomposition and nutrient cycling in coastal forest ecosystems. *Can. J. For. Res.* 16:983–989.

Springett, J.A. and J.K. Syers. 1979. The effect of earthworm casts on ryegrass seedlings. p. 44–47. In: Crosby, T.K., and R.P. Pottinger (eds.) *Proc. 2nd Australasian Conf. Grassland Invert. Ecol.* Government Printer, Wellington, New Zealand.

Stockdill, S.M.J. 1982. Effects of introduced earthworms on the productivity of New Zealand pastures. *Pedobiologia* 24:29–35.

Svensson, B.H., U. Boström, and L. Klemedston. 1986. Potential for higher rates of denitrification in earthworm casts than in the surrounding soil. *Biol. Fertil. Soils* 2:147–149.

Syers, J. K., A.N. Sharpley, and D.R. Keeney. 1979. Cycling of nitrogen by surface-casting earthworms in a pasture ecosystem. *Soil Biol. Biochem.* 11:181–185.

Tiwari, S.C., B.K. Tiwari, and R.R. Mishra. 1989. Microbial populations, enzyme activities and nitrogen-phosphorus-potassium enrichment in earthworm casts and in the surrounding soil of a pineapple plantation. *Biol. Fertil. Soils* 8:178–182.

Tomati, U., A. Grappelli, and E. Galli. 1988. The hormone-like effect of earthworm casts on plant growth. *Biol. Fertil. Soils* 5:288–294.

Trigo, D. and P. Lavelle. 1993. Changes in respiration rate and some physicochemical properties of soil during gut transit through *Allolobophora molleri* (Lumbricidae, Oligochaeta). *Biol. Fertil. Soils* 15:185–188.

West, L.T., P.F. Hendrix, and R.R. Bruce. 1991. Micromorphic observations of soil alteration by earthworms. *Agric. Ecosystems Environ.* 34:363–370.

Wolters, V. and R.G. Joergensen. 1992. Microbial carbon turnover in beech forest soils worked by *Aporrectodea caliginosa* (Savigny) (Oligochaeta: Lumbricidae). *Soil Biol. Biochem.* 24:171–177.

Yeates, G. W. 1981. Soil nematode populations depressed in the presence of earthworms. *Pedobiologia* 22:191–195.

Zachmann, J.E. and D.R. Linden. 1989. Earthworm effects on corn residue breakdown and infiltration. *Soil Sci. Soc. Am. J.* 53:1846–1849.

Earthworms and Their Influence on Soil Structure and Infiltration

A.D. Tomlin, M.J. Shipitalo, W.M. Edwards, and R. Protz

I. Introduction

Edwards (1980) wrote that the consensus of evidence was that earthworms improve soil structure, fertility, organic matter decomposition, aeration and drainage. The processes by which these improvements occur are still open questions, especially their contribution to formation and maintenance of soil structure. The questions for soil scientists include: (1) What are the direct and indirect effects on soil structure at the microfabric scale? (2) What changes occur to water and air infiltration in soil? (3) What is the rate of turnover of organic and mineral matter? and (4) What is the contribution to soil aggregate stability? Additionally, questions arise with regard to the methods required for

1-56670-053-1/95/$0.00+$.50

characterizing biogenic soil structure, procedures for managing earthworm populations to increase their beneficial effects, and identifying the critical experiments needed to answer these questions.

Darwin's monograph (1881) was the first systematically documented attempt to relate earthworm activity with their effects on soil processes. He clearly understood that earthworms could, with time, bury large surface stones and features, and transport surface litter into soil via their casting activity. Darwin's descriptions were remarkably detailed at the scale of what today we would refer to as peds and landscapes. He realized that what is now called soil structure (in 1881, there was no formal notion of soil structure) is affected by the action of earthworms, for he says of them "...their chief work is to sift the finer from the coarser particles, to mingle the whole with vegetable debris, and to saturate it with their intestinal secretions" (Darwin 1881, pp. 174–175). This description is close to that of Lee (1985) who described the earthworm burrow lining as composed of illuviated material, dissolved materials, and earthworm secretions and excretions.

Bal (1974, 1982) demonstrated the contribution of faunal activity and soil structure by using micromorphological techniques to identify the characteristic features of fecal material from earthworms and other soil fauna. Altemuller and co-workers (Altemuller 1974, Altemuller and van Vliet-Lanoe 1990, Altemuller and Joschko 1992) used fluorescent staining techniques in combination with micromorphological methods to illustrate the spatial distribution of soil biota in the soil.

Different species of earthworms have different ecological strategies and behaviors (Bouché 1977) that take them to different compartments of the soil ecosystem. Anecic earthworms such as *Lumbricus terrestris* L. feed at the surface and transport surface litter vertically into the soil profile. Endogeic species such as *Aporrectodea turgida* Savigny are geophagous, obtaining their food by filtering it from the large amounts of soil that they pass as they burrow horizontally through surficial soil layers. Different earthworm behavior patterns and preferences for organic matter sources as food (Hendrikson 1990) have significant consequences for soil structure at the microfabric scale.

The production of earthworm channels by their burrowing habit has attracted attention of soil physicists because of the probability that this activity would strongly affect air and water infiltration into the soil (Ehlers 1975). An associated concern is changes in agrochemical leaching rates due to earthworm burrowing (Baker 1987, Shipitalo et al. 1990, Stehouwer et al. 1993). In some instances continuous no-tillage (NT) sites exhibit decreased surface run-off and increased infiltration and leaching relative to conventionally tilled (CT) sites (Edwards et al. 1988b). The issue is further complicated by earthworm burrow lining material which differs chemically (Tomlin et al. 1993) from surrounding matrix soil, and may influence solute transport through the burrows. Stehouwer et al. (1993) demonstrated that atrazine sorption and retention was up to three times greater on *L. terrestris* burrow walls than in an unlined void, and that most of the difference in sorptive capacity was explained by organic carbon in the burrow linings.

This review covers three major topics, (1) earthworm influences on soil structure and infiltration, (2) mechanisms for aggregate (cast) stabilization, and (3) new approaches for characterizing earthworm cast stability.

II.   Earthworm Influences On Soil Structure

A.   Effects on Porosity

The readily measurable effects that earthworms have on porosity in soils are created by burrowing or surface and subsurface casting (Satchell 1967, Edwards and Lofty 1977). In tilled soils, human activity may create changes that temporarily mask the effects of earthworms on porosity. However, in forests and grasslands, earthworm numbers have sometimes been correlated with higher porosity (Hoeksema and Jongerius 1959, Satchell 1967). Several authors report that introduction of earthworms to areas with no resident populations resulted in small increases in total porosity (Edwards and Lofty 1977, Springett et al. 1992). Knight et al. (1992), however, reported that burrowing activity by introduced earthworms increased macroporosity in pastures.

No-tillage, unlike conventional practices, has been positively correlated with higher earthworm densities (Edwards and Lofty 1982, Mackay and Kladivko 1985) and higher burrow density (Lal 1974, Ehlers 1975, Edwards and Lofty 1982). In the early 1970's, several scientists counted earthworm burrow densities in the field (Lee 1985), and related number and size of burrows to air and water movement. Kretzschmar (1987) found as many as 10,000 *Allolobophora* spp. burrows per square meter, and Edwards et al. (1988a) reported 1.6 million *L. terrestris* burrows ($\geq$5 mm in diameter) per hectare.

Geophagous earthworms generate burrows (amongst the largest pores in many soils) with a diameter similar to their body diameter. The effect of earthworm burrowing on total porosity may depend on the bulk density of the soil horizon before earthworm activity, and whether most of the casting is deposited on the soil surface or in the same horizon from which it originated.

Although earthworm burrows may account for a small fraction of soil volume (Gantzer and Blake 1978, Lauren et al. 1988), burrow continuity is an important attribute of earthworm induced porosity (Lee 1985). Although uneven introductions of European earthworms to the U. S. over the past three centuries (Reynolds 1977) may account for population differences in similarly managed fields, soil management methods can influence earthworm densities (Gantzer and Blake 1978, Shipitalo and Protz 1987). It may, therefore, be possible and advantageous to manage the number, diameter, and length of earthworm burrows in field situations to influence air, water, and chemical movement (Smettem 1992) (see Lee this volume).

The continuity and shape of the burrow system and its access to water influence the effectiveness of burrow infiltration. Teotia et al. (1950) poured molten lead into worm holes and washed away the soil to reveal the continuity and tortuosity of the burrow system. Recently, burrow length, orientation, and branching frequency were measured during careful excavation by McKenzie and Dexter (1993).

B.   Earthworm Burrows and Infiltration

The tremendous capacity of earthworm burrows for infiltration has been directly demonstrated by Bouché (1971), Ehlers (1975), and Bouma et al. (1982). Smettem and Collis-George (1985b) showed that a single, continuous 0.3 mm diameter macropore in a 100-mm diameter soil column can conduct more water than the rest of the column. Methods for evaluating porosity, including the effects of biopores on transport of water and solute, have been reviewed by Beven and Germann (1982), Nielsen et al. (1986), Wagenet (1986), Brusseau and Rao (1990), Coltman et al. (1991), and Edwards et al. (1993a).

The open, vertical burrows of *L. terrestris* have been implicated in "short-circuiting" or "by-pass flow" (Bouma et al. 1981, Bouma et al. 1983, Bouma 1991). When earthworms were introduced to soil columns that had been clogged with septic tank effluent, their burrowing activity reopened the columns to flow rates greater than initial values (Jones et al. 1993). Colored dyes and other tracers have been used by investigators to confirm that flow in burrows and similar continuous macropores occurs under a wide range of conditions (Douglas et al. 1980, Tyler and Thomas 1981, Germann et al. 1984, Smettem and Collis-George 1985a, Smith et al. 1985, Everts et al. 1989, Zachmann and Linden 1989, Andreini and Steenhuis 1990).

All burrows, however, do not conduct water during all storm events (Shipitalo et al. 1990, Trojan and Linden 1992), but adding earthworms to worm-free soils usually results in increased infiltration rates (Kladivko et al. 1986). Previous soil moisture content influences flow in earthworm burrows as does rainfall intensity (Edwards et al. 1992). Frequently, only a few of the apparently open burrows conduct water ahead of the wetting front, and those few "good conductors" are active in successive storms (Edwards et al. 1992, Trojan and Linden 1992). Effective conducting burrows may be made by both anecic and endogeic species (Joschko et al. 1992).

C.   Chemical Movement in Earthworm Burrows

Chemical transport in earthworm burrows has been documented for several years, but only recently have conditions under which such movement is important been given much consideration. Because water may preferentially infiltrate earthworm burrows and bypass adjacent soil, water and chemicals that move in the burrows can be of very different quality than in the soil matrix. For example, if burrow flow occurs in nearly vertical *L. terrestris* burrows as the result of high-intensity rain storms immediately after the surface application of chemicals, entrained chemicals can be quickly carried through much of the soil without attenuation. In much of the central U. S., fertilizers and pesticides are usually surface-applied at the time of year that storms are strongest. However, after rainfall has moved the surface-applied chemicals into the topsoil, by-pass flow may be carrying relatively clean rain water past the chemical-laden soil water held in the matrix. In the latter case, flow in the earthworm burrows may be of better quality than that held or moving down in the soil matrix.

Edwards et al. (1988a, 1989) characterized earthworm burrows in long-term no-till cornfields and documented the hydrologic conditions under which nitrate infiltrated in *L. terrestris* burrows. In English pastures, moderate population densities of *L. rubellus* and *A. caliginosa* increased nitrate transport in leachate three-fold (Knight et al. 1992). Depending upon weather conditions, which influence fertilizer uptake, nitrification, and denitrification, nitrate concentrations in burrow flow can be high at any time of the year (Shipitalo et al. 1994). Bromide, a highly soluble ion, is used as a non-reactive tracer to simulate nitrate movement in transport studies. Like nitrate, bromide movement in burrows near the surface and transport below the plow layer can be greatest whenever rapid water movement is enhanced by the introduction of earthworms (Zachmann et al. 1987).

In contrast, pesticide concentrations in burrow flow are highest in the first storm following surface application. Both concentration and transport decrease in subsequent storms (Shipitalo et al. 1990) and with time between application and the first following storm (Edwards et al. 1993b). A small storm event that does not induce burrow flow reduces herbicide concentration and transport in burrow flow of subsequent storms (Shipitalo et al. 1990). Because high rainfall intensities increase the flow of water in burrows, they also increase transport of available surface-applied chemicals (Edwards et al. 1992).

Chemical composition of the burrow lining may influence solute transport through the burrows (Stehouwer et al. 1993). Distilled water poured into the top of *L. terrestris* burrows may pick up nitrate from the drilosphere, whereas herbicide concentrations may decrease during burrow infiltration (Edwards et al. 1992, Stehouwer et al. 1994). Stehouwer et al. (1993) separated burrow linings from the surrounding soil to identify the lining properties that could increase herbicide sorption and retard herbicide transport into and out of (through) the burrows.

D.   Other Effects of Earthworms on Soil Structure

At a long term experiment near Guelph, Ontario, the addition of three different types of municipally-generated sludge to the soil surface for 8 years, and continuous bromegrass for 18 years created a 12-cm thick A_h horizon with significantly lower bulk density, and, increased porosity. The results of this experiment of anthropogenically-initiated faunal amelioration of soils are summarized in Figure 1: (1) soil elevation was slightly increased, (2) a new A_h horizon encroached on the A_p and A_e horizons, (3) the A horizon darkened because of the incorporation of organic matter, and (4) a 1-mm thick surface sludge application over a 10 x 10 cm surface was distributed as a coating on the walls of an earthworm burrow 50-cm long. Sludge application induced a large increase in populations of *L. terrestris* in the treated soils because organic matter, nutrients, and water were no longer limiting. This meant that turnover of macronutrients and physical mixing of organic matter and mineral matrix

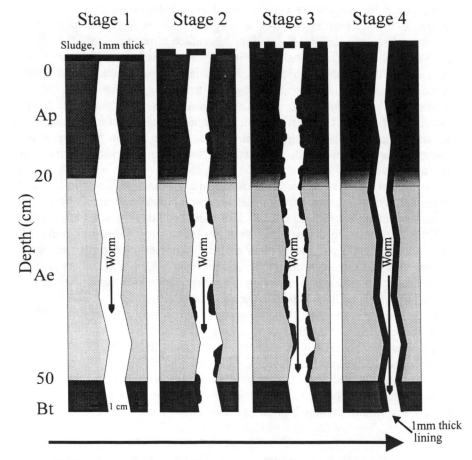

Figure 1. Proposed scheme by which earthworms transport surface applied sewage into their burrows.

were greatly accelerated by earthworms during the course of the 16 years of the experiment's duration.

E.   Ingestion and Casting Activity

Earthworms are believed to be a major influence on soil aggregation because of the amounts of organic matter and soil they ingest, but quantification is difficult. The difficulty arises because measuring the size and composition of earthworm populations, their feeding and casting activities, and correlating that with type of food source and population structure under field conditions are challenging.

Field studies suggest that earthworms ingest a considerable amount of organic debris and soil. Estimates of cast production are quite variable ranging from 1.5 to 2600 Mg ha^{-1} yr^{-1} (Watanabe and Ruaysoongnern 1984) (note: 13 Mg represents a 1 mm thick layer of soil/ha at a bulk density of 1.3 Mg m^{-3}). Guild (1955) calculates that earthworms can ingest 22 to 27 Mg ha^{-1} yr^{-1} of cow dung, and Satchell (1967) calculates that earthworms in a deciduous forest consume an annual leaf fall of 3 Mg ha^{-1} in approximately 3 months. Extremely high rates of cast production are most common in tropical soils, but all estimates for temperate soils are <100 Mg ha^{-1} yr^{-1}. It should be noted that extreme values reported for tropical areas are from old studies and may be questionable. Recently, however, Lavelle et al. (1989) reported a cast production rate of 1200 Mg ha^{-1} yr^{-1} in the Ivory Coast.

Nevertheless, based on available data, it is obvious that earthworms in both tropical and temperate region soils can process substantial amounts of mineral and organic material annually. Moreover, it is recognized that field estimates of casting activity are probably gross underestimates of actual amounts ingested because many species do not commonly cast onto the soil surface, and sub-surface casting is often ignored (Evans 1948, Edwards and Lofty 1977 p. 145, Bal 1982 p. 12). It is logical to assume, therefore, that changes in population size, composition, or activity due to changes in cultural practices may have an effect on casting activity and soil aggregation.

Laboratory studies of feeding and casting activity overcome some of the problems inherent in field studies by controlling earthworm populations and environmental conditions. The results of laboratory studies suggest that food ingestion rate is influenced by food source (Van Rhee 1963), food supply (Raw 1962), age and chemical composition of the food (Edwards and Heath 1975), temperature (Knollenberg et al. 1985), moisture (Satchell 1967 p. 269), and earthworm species, body size, and maturity (Barley 1959a). Casting activity, in turn, is a function of feeding activity. Abbott and Parker (1981) and Martin (1982) suggest that earthworms ingest more soil in an attempt to obtain sufficient food when supply is limited, and consequently cast more frequently. Martin found, however, that some species ingest the most soil when intermediate levels of food are supplied. Under experimental conditions, Barley (1959a) and Parle (1963a) found that earthworms cast more when first placed in soil in pots, because they are establishing new burrows and are not feeding. Abbott and Parker (1981) found that food source and food placement, as well as earthworm species, influence the amount of surface casts produced. They found that incorporating clover hay into soil results in more surface cast production by *Eisenia foetida* (Savigny) than when clover is left as a mulch on the soil surface.

Other factors affecting casting activity are soil temperature (Hartenstein et al. 1981, Joannes and Kretzschmar 1983), and soil pH and Ca levels (Springett and Syers 1984, Nielson 1951). Surface casting activity is reportedly encouraged by high soil bulk densities (Guild 1955, Thomson and Davies 1974) and influenced by soil texture (Teotia et al. 1950, Thomson and Davies 1974).

A variety of factors influence earthworm feeding and casting activity and these factors are not independent, perhaps accounting for some contradictions among studies. Ideally, it would be useful to develop a model which incorporates the key factors affecting food ingestion and casting along with measurements of field populations to predict activity in the field. However, we are a long way from achieving this goal. As a starting point, Shipitalo et al. (1988) used laboratory earthworm activity data in conjunction with field population data to calculate hypothetical food consumption and cast production rates in order to estimate the effects of food supply and food quality on earthworm populations and cast production. James (1991) estimated total cast production of mature *Diplocardia* populations in a tall grass prairie using cast production-temperature relationships obtained in the laboratory, combined with field population density estimates and soil climate data. Such approaches could be extrapolated to help us understand the changes in earthworm populations and activity that occur when land management practices are altered.

Manuring cultivated fields increases earthworm populations (Berry and Karlen 1993), and minimum tillage practices usually result in increased populations and biomass of earthworms (Edwards and Lofty 1982, De St. Remy and Daynard 1982, Berry and Karlen 1993). Crop selection and rotation have significant effects on subsequent earthworm populations with continuous soybean cropping reducing earthworm numbers the most, presumably due to the reduction in soil organic matter associated with this practice (Tomlin et al. 1993). All of these changes in earthworm populations induced by agronomic practice suggest that there could be some optimum combination of cropping and tillage practices that could be implemented to maximize earthworm numbers even to the point of manipulating species within a field that would enhance or remediate soil structure (see Edwards et al. and Lee, this volume).

III. Research Issues on Stability of Earthworm Casts

A. Methodological Issues Related to Measuring Stability

Just as the amount of material processed by earthworms is poorly quantified, so too is the aggregate stability of the food-soil mixture egested by earthworm as casts. Numerous studies assessed the stability of casts (Table 1), but we have little quantitative information on how important earthworm casting activity is in promoting soil aggregation in a natural setting.

There are a number of methods used to quantify aggregate stability and results obtained are technique dependent. Additionally, there is growing acceptance that there are different levels of organization within aggregates, and the major binding agents can differ among these levels. Edwards and Bremner (1967) first suggested that microaggregates (<250 μm) bond by different mechanisms than macroaggregates (>250 μm). Tisdall and Oades (1982) refined this concept and proposed a model that engenders several levels of aggregation to account for aggregate stability in Australian red brown earths. In their model, microaggregates are bonded by persistent, degraded, aromatic humic materials, and transient to persistent polysaccharides. Temporary binding agents, such as roots and fungal hyphae, bind microaggregates together into macroaggregates. They suggest that the model is appropriate for most soils where organic matter is the major binding agent, but the levels of aggregation may depend on soil type. Thus, different methods for measuring aggregate stability may assess aggregation at different levels, explaining why various techniques seldom yield results that are directly comparable.

Moisture status can also affect the measured stability of aggregates. Reid and Goss (1981) noted that drying of soil samples prior to assessing aggregate water stability can markedly affect measured stabilities and that the magnitude of the drying effect can be treatment dependent. Utomo and Dexter (1982) demonstrated that wetting and drying cycles and thixotropic hardening affect aggregate stability. Furthermore, Beare and Bruce (1993) have shown that the method by which samples are wetted before being subjected to wet sieving can affect the measured stabilities and the level of aggregation (micro vs. macro) being assessed. In order to avoid the effects of air drying and storage, especially when biologically mediated effects on soil structure are being investigated, they recommend that samples be analyzed immediately after sampling and that gentle pre-wetting be used. Additionally, they wisely suggest that complete specifications of the procedures used accompany aggregate stability analyses.

A further complication related to assessing the stability of earthworm casts has to do with choosing a basis by which to compare changes in stability. In field studies, casts are most often compared to soil collected nearby, which is usually coarser in texture. This can be attributed to (1) comminution of the mineral particles ingested, (2) selective avoidance by the earthworms of larger particles, and (3) ingestion of finer-texture material from elsewhere in the profile. Although some research work suggests that comminution of mineral

Table 1. Characteristics of studies in which the stability of earthworm casts was assessed.

Reference	Field/Lab Study	Location	Species	Method used to Measure Stability
Barley 1959b	Lab	Adelaide, Australia	*Allolobophora caliginosa* (Sav.)	Wet Sieving
Dawson 1947	Lab	Maryland, USA	*Lumbricus terrestris* (NS)	Falling Water Drop
Dutt 1948	Both	New York, USA	NS	Wet Sieving
Guild 1955	Lab	Edinburgh, UK	*Lumbricus rubellus* (Hoff.) *Lumbricus terrestris* (L.) *Allolobophora longa* (NS) *Allolobophora caliginosa* (Sav.) *Dendrobaena subrubicunda* (Eisen)	Dry and Wet Sieving
Lal and Akinremi 1983	Field	Ibadan, Nigeria	*Hyperiodrilus* spp.	Wet Sieving
Low 1972	Field	Derbyshire, UK	NS	NS
Marinissen and Dexter 1990	Lab	Netherlands and South Australia	*Aporrectodea caliginosa* (NS)	Dispersible Clay
McKenzie and Dexter 1987	Lab	Glen Osmond, South Australia	*Aporrectodea rosea* (NS)	Crushing—Tensile Strength
Parle 1963b	Field	Rothamsted, UK	*Allolobophora terrestris* (Sav.) forma *longa* (Ude)	Permeability
Shipitalo and Protz 1988	Lab	Ontario, Canada	*Lumbricus rubellus* (Hoff.) *Lumbricus terrestris* (L.)	Dispersible Clay
Swaby 1950	Field	Rothamsted, UK	*Allolobophora nocturna* (NS)	Wet Sieving
Teotia et al. 1950	Both	Nebraska, USA	*Helodrilus caliginosus* (Sav.) *Helodrilus parvus* (Eisen) *Octolasium lacteum* (Orley) *Diplocardia riparia* (Smith)	Falling Water Drop
West et al. 1991	Lab	Georgia, USA	*Lumbricus rubellus* (NS)	Wet Sieving

NS — not specified in the reference.

particles occurs during passage through the gut of an earthworm (Edwards and Lofty 1977), significant comminution of mineral grains during a single passage is unlikely with most materials, and they seriously question whether mineral comminution occurs. It is logical to assume that there is an upper limit to the size of mineral grains that earthworms can ingest and this size limit depends on the sizes and species of worm being investigated. Similarly, it is probable that earthworms mix material from several horizons encountered while burrowing.

Because of selective ingestion and mixing of material from elsewhere in the soil, ingested material may have different chemical and physical properties than the soil to which it is compared. Thus, it can be difficult to separate these effects from those directly attributable to the actions of the earthworms on the material ingested. This is important when the factors and mechanisms affecting aggregate stabilization in casts are being investigated. Although there is no simple way to account for this effect in the field, its potential importance must be recognized. In laboratory studies, where earthworm are confined to a uniform soil material, the mixing effect is eliminated and procedures which account for the effects of selective ingestion on aggregate stability can be used (Shipitalo and Protz 1988).

Despite the problems in quantifying the stability of casts, the existing body of research suggests that a number of factors can have an effect on stability of earthworm-formed aggregates. Both Parle (1963b) and Teotia et al. (1950) indicate that cast stability increases with age, the maximum stability being attained 15 days after excretion. Marinissen and Dexter (1990) also found that cast stability increased with age when measured up to 42 days after excretion. Conversely, Dawson (1947) reported that soil aggregates removed from the intestines of earthworms are more stable than casts and concluded that aging and post-depositional microbial activity resulted in the destruction of bonding agents and a decline in aggregate stability. Hopp and Hopkins (1946) note that aging up to four weeks caused slight decreases in the stability of casts. Shipitalo and Protz (1988) found that the effect of age was dependent on the diet provided to the worms and whether casts were dried before analysis. Unless the earthworms were provided a food source which they accepted, aging up to 32 days had no effect on cast stability.

If casts were aged moist but allowed to air-dry prior to analysis, cast age had no effect. As Shipitalo and Protz (1988, 1989) and Marinissen and Dexter (1990) demonstrated, drying can override the effect of aging on measured cast stabilities. When samples were kept moist, freshly formed casts were invariably less stable than uningested, moist, soil. It is now recognized that earthworms disrupt existing aggregates in the process of forming new ones, and restoration or improvement of aggregate strength only occurs with aging or drying (Barois et al. 1993, Lee and Foster 1991, Marinissen and Dexter 1990, McKenzie and Dexter 1987, Shipitalo and Protz 1988, 1989). Thus, for a finite period of time, casts are less stable than the soil from which they were derived. This may, in part, account for the observation that surface casting activity by earthworms contributes to soil erosion under some conditions (Madge 1969, Sharpley et al.

1979). Under some conditions wetting and drying cycles may be an important stabilization process, but air drying does not normally occur below the upper few centimeters of soil in temperate regions (Reid and Goss 1981). While wetting and drying cycles may be important for stability of surface casts, the process is probably of limited significance for subsurface casts.

Experimental evidence on the importance of food source in aggregate stabilization by earthworms is conflicting. Dawson (1947) reports aggregate stabilization in the absence of added food and that addition of lespedeza hay only slightly increased cast stability, but Swaby (1950) found that air-dried casts were more water stable than soil only when they contained "nutritive" organic matter. Teotia et al. (1950) found that worm casts were most water stable when collected from pots to which alfalfa mulch was added, less stable when from pots to which straw mulch was added, and least stable when no mulch was added. Shipitalo and Protz (1988) found that food source type provided to worms and the amount of residual organic matter remaining in casts both affect aggregate stability. In turn, the amount of organic matter incorporated into an individual cast was dependent on both food source and whether the worms were actively feeding or burrowing only. In trials where no food was provided to the worms, organic carbon originating from the worms as digestive secretions appeared to be effective in stabilizing casts, but since the amounts were small in comparison with treatments where food was provided, the effects on cast stability were minimal (Shipitalo and Protz 1988).

Earthworm species and soil texture have been shown to influence the water stability of casts. Teotia et al. (1950) found that casts produced under straw mulch by *Helodrilus calignosus* were most stable and those of *Diplocardia riparia* least stable of four species they investigated, and that the degree of improvement in cast stability increased as soil texture became finer. Guild (1955) reported method-dependent differences in aggregate stability in soils worked by different earthworm species. Guild observed a 16 to 20%, and up to a 40%, improvement in soil aggregation due to earthworm activity for *L. rubellus* and *L. terrestris*, respectively, when assessed by the dry shaking and sieving method. When measured by wet sieving, Guild obtained a 20% and a 50% improvement for *L. rubellus* and *L. terrestris*, respectively. Shipitalo and Protz (1988) also investigated the effects of these two species on soil structure. They found that casts produced by *L. rubellus* were generally more water stable than those produced by *L. terrestris* when provided the same food source. The greater stability of casts produced by *L. rubellus* was attributed to more organic matter incorporation into their casts than those of *L. terrestris*, and not to any difference in the way these species affected soil aggregation. In a study where the interaction among species was investigated, Shaw and Pawluk (1986) found that development of soil structure was most pronounced when both anecic and endogenic species were introduced into microcosms. Aggregate stability, however, was not assessed in this study.

B.   Concepts and Theories for Modeling Earthworm Cast Stability

Based on the factors which affect cast stability and ancillary data collected by investigators, numerous theories on the nature of the bonding substances and mechanisms in earthworm casts have been proposed, including:

- stabilization by internal secretions of earthworms (Dawson 1947)
- mechanical stabilization by plant fibers incorporated into casts (Dawson 1947, Lee and Foster 1991)
- mechanical stabilization by fungal hyphae (Parle 1963b, Marinissen and Dexter 1990, Lee and Foster 1991)
- stabilization by bacterial gums (Swaby 1950)
- stabilization via formation of organo-mineral bonds in the form of calcium humate (Meyer 1943 as cited by Satchell 1967, p. 294) or mucilage (Dutt 1948)
- stabilization due to wetting and drying cycles with (Shipitalo and Protz 1988, 1989) or without organic bonding (Marinissen and Dexter 1990)
- age-hardening/thixotropic effects combined with organic bonding (Shipitalo and Protz 1988, 1989).

Each of these theories probably has some validity under the conditions in which the experiments were conducted. For instance, Shipitalo and Protz (1989) suggested that drying stabilized casts due to dehydration and irreversible bonding of incorporated organic matter, whereas Marinissen and Dexter (1990) indicated that organic bonding materials did not play a major role in stabilizing dried casts. This finding must be interpreted in light of the fact that Shipitalo and Protz (1989) provided a food source to the worms, whereas no food was provided to the worms in the experiment conducted by Marinissen and Dexter (1990). The degree of aggregate stabilization detected is dependent on the properties of the soil and organic matter ingested by earthworms, the procedure by which aggregation is measured, and the level of aggregation that a particular technique assesses. None of the theories is mutually exclusive and, in all probability, more than one mechanism contributes to aggregate stabilization under field conditions. What is important is whether a particular mechanism plays a dominant role under the given set of circumstances.

Most theories assign a significant role to organic compounds as the bonding agent responsible for stabilizing the aggregates in casts. The nature of these organic compounds and the type of bonding that occurs is probably highly dependent on the chemical and physical characteristics of the food and soil ingested by the worms, and on the amount of reprocessing that occurs in the gut and subsequently in the deposited casts. Apart from these factors, it is now generally recognized that the distribution of the active fraction of the organic matter in the soil can be more important than the total amount present. For this reason a number of recent studies have utilized micromorphological techniques to investigate the distribution of organic matter in casts (Shaw and Pawluk 1986, Shipitalo and Protz 1989, Lee and Foster 1990, West et al. 1991, Altemuller and Joschko 1992, Barois et al. 1993). The general consensus derived from these studies is that a portion of the ingested organic matter can become highly fragmented and partially humified and serve as foci for aggregate reformation. The organic substances most often implicated as bonding agents are polysaccharides and mucopolysaccharides (Shaw and Pawluk 1986, Shipitalo and Protz 1989, Lee and Foster 1991, Barois et al. 1993) but the specific composition of these substances has not been established. Based on the use of selective chemical extractants and measurement of subsequent losses in aggregate stability, Shipitalo and Protz (1989) concluded that bonding was primarily the result of Clay-Polyvalent cation-Organic Matter linkages (i.e., C-P-OM bonding of Edwards and Bremner 1967) with the materials investigated. Additionally, the type of bond formed (i.e., water bridges, cation bridges, coordination complexes) was dependent on the food source provided to the worms, and age and moisture status of the casts.

C.   Earthworm Casts and Soil Structure

Only two classes of soil structure (Bullock et al. 1985, Fitzpatrick 1986), granules or crumbs (Figure 2), can be viewed as being formed dominantly by biotic processes. Individual earthworm fecal pellets could be viewed as crumbs, and casts (multiple pellets) could be viewed as granules. This scheme would fit into the hierarchy of aggregates developed by Oades and Waters (1991) and Oades (1993) (Figure 3).

The search continues for an understanding of the processes which must occur to make stable aggregates. Obviously, passage of organic material and mixing of mineral particles with a varying bacterial suite in the gut of an earthworm is part of the aggregate formation and stabilization process. Pedersen and Hendriksen (1993) concluded from data on individual bacterial concentrations that some bacteria decrease in numbers in the foregut, some bacteria increase during, and some bacteria are unaffected by passage through the gut. Earthworms exhibited selective bacterial feeding patterns that can lead to changes in bacterial survival. Stephens et al. (1993) demonstrated that the earthworm, *A. trapezoides*, in numbers comparable to field densities, transported significant amounts of the bacteria, *Pseudomonas corrugata* (strain 2140-R) from the soil surface to a depth of 9 cm in laboratory pot experiments in 9 days. This clearly demonstrates that earthworms mix organic matter and soil mineral particles together.

Heine and Larink (1993) concluded that sand and/or inorganic particles, nitrogen content and particle size of organic matter are an essential influence on the nutrition of *L. terrestris*. This would make earthworms central to soil aggregate formation. Once aggregates are formed and organic acids and enzymes interact with mineral particles, certain parts of the organic fraction are stabilized and protected by the clay and silt fractions (Figure 4, Tisdall and Oades 1982, Hassink et al. 1993). If an earthworm sac (fecal pellet) stays intact during a drying period, the bacterial decomposition products will be drawn into the silt

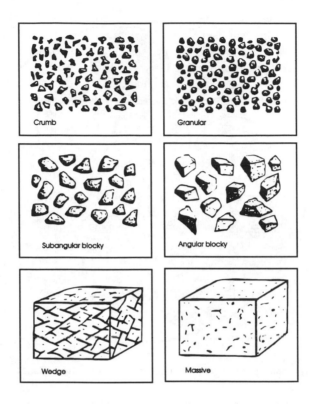

Figure 2. Diagrammatic representation of some types of soil structure (not to scale). Modified from Fitzpatrick (1986).

and clay particle interspaces as in the last stage presented by Tisdall and Oades (1982). This stable aggregate may then be reingested several times, followed by drying, resulting in the protection of older organic matter. Thus, when the surface organic matter is dated it will be modern, but internal organic matter will be older. Skjemstad et al. (1993) measured this organic matter as 200 to 320 years old. In other soils the organic matter may be older depending on the nature of soil processes forming the aggregates. Age dating will allow us to estimate the rates at which degraded A_p horizons can be rehabilitated.

Wershaw (1992) reviewed concepts of biomass degradation in soils and sediments, and proposed that depolymerization and oxidation reactions that occur during enzymatic degradation of biopolymers produce amphiphiles, molecules that have polar (hydrophilic) and nonpolar (hydrophobic) moieties. Amphiphiles resulting from partial oxidative degradation of biomass assemble spontaneously into ordered structures in which the hydrophobic moiety forms the interior of organic molecule aggregates, and the hydrophilic moiety forms the exterior surface. Amphiphiles can be viewed as micelles in solution, and at a larger, older stage of aggregation as vesicles (Figure 5). The humus-ordered aggregates in soils likely exist as bilayer membranes (Figure 5) coating mineral grains.

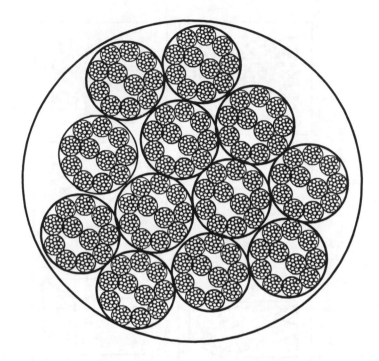

Figure 3. The concept of aggregate hierarchy. Modified from Oades (1993).

These ordered aggregates of amphiphilic molecules constitute the humus in soils. Wershaw's view of a progressive accumulation of organic molecules through the micelle, vesicle, and bilayer lipid membrane stages towards the protection of mineral particles parallels Tisdall and Oades' concepts (Figure 4). And earthworms accelerate the rate of association of organic molecules with mineral particles (Stephens et al. 1993), thus enhancing the formation of crumb and granular structure (Pedersen and Hendriksen 1993).

IV.   Research Imperatives

A.   Solute Transport

Effects of earthworm activity on soil porosity and water infiltration have been well documented. But we are only beginning to understand the conditions under which earthworm burrows transport solutes, particularly surface-applied chemicals, into deeper soil or groundwater. Controlled experiments under field conditions in addition to laboratory process studies are needed to address this problem.

Major binding agent

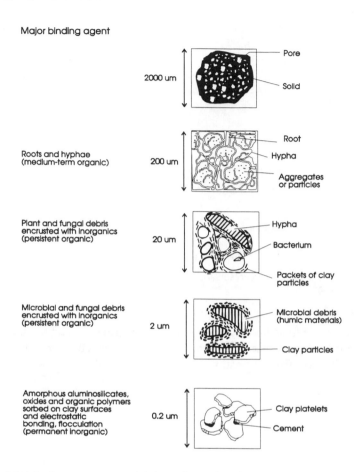

Roots and hyphae
(medium-term organic)

Plant and fungal debris
encrusted with inorganics
(persistent organic)

Microbial and fungal debris
encrusted with inorganics
(persistent organic)

Amorphous aluminosilicates,
oxides and organic polymers
sorbed on clay surfaces
and electrostatic
bonding, flocculation
(permanent inorganic)

Figure 4. Model of aggregate organization with major binding agents indicated. Modified from Tisdall and Oades (1982).

B.   Properties of Burrow Linings

Earthworm burrows also may function both as a source and a sink for certain compounds in water flowing through them. Research is needed to better characterize the physical, chemical, and microbiological nature of burrow linings, particularly their exchange properties and organic chemical composition.

C.   Soil Processing and Soil Structure

We have no reliable data as to how much material is processed by earthworms and, given the difficulty associated with measuring this in the field, this will likely remain a problem. Rather than let this deficiency impede progress, we

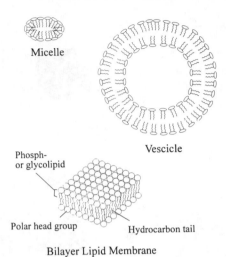

Micelle

Vescicle

Phosph-
or glycolipid

Polar head group   Hydrocarbon tail

Bilayer Lipid Membrane

Figure 5. Diagrammatic representations of amphiphilic molecules.   Modified from Wershaw (1992).

should continue to address the question of how earthworms affect soil structure, because it is obvious that earthworms process large amounts of soil and organic matter that affect soil aggregation.

D.   Aggregate Stabilization

We now recognize that earthworm casting activity initially destabilizes aggregation, but net improvements in aggregate stability can occur over time. Stabilization is influenced by the amount and nature of organic matter incorporated, cast age and moisture status. What is unknown is whether these improvements in aggregate stability measured in the laboratory, or on individual casts collected in the field, translate into a net improvement in soil structure in the field (and how long any improvement persists). Under natural conditions soil aggregation is a dynamic process. Aggregate formation and destruction can occur simultaneously, and improvement in soil structure occurs when the former predominates. We know that soils with good structure and stable aggregates exist where earthworms are absent. We need to know what properties enhance aggregate stability in these soils and, conversely, in soils without these properties, how earthworm activity may impact aggregate stability. The key probably lies in the nature of organic compounds and particularly in the rate of their association with mineral particles, which is accelerated by earthworm activity.

E.   Field and Laboratory Studies

In the field there are innumerable interactions among anecic and endogeic earthworms, and among earthworms and other soil-inhabiting animals, such as enchytraeids and microarthropods. Given the complexity of the soil ecosystem, perhaps the best approach to answering our questions is to study intact systems where earthworm activity has been experimentally manipulated by selectively defaunating or selecting soils devoid of earthworms, introducing earthworms or stimulating earthworm activity by altering fertility, pH, organic matter, or calcium levels (e.g., Hamilton and Dindal 1989). These studies must be long-term in order to factor out natural variations inherent in field studies and should be conducted in conjunction with controlled laboratory studies that are process-oriented.

F.   Organomineral Interactions and Soil Rehabilitation

Finally, we need experiments where earthworms are fed a single plant material within a mono-mineralic soil (a pure clay mineral) to clarify the process of organic molecule-clay mineral interaction within earthworm fecal pellets, and the subsequent bacterial decomposition processes. The exact orientation of mineral grains in fine aggregates may be measured by high resolution electron microprobe X-ray backscatter images (Petruk 1989) using image analysis software such as Mocha® (Jandel Scientific 1993). In consequence, we should be able to relate the crumb/granular soil structure of different A_h horizons to soil texture and faunal activity. The rate of earthworm and associated faunal activity in never-cultivated soils could be measured using gamma isotopes (McCabe et al. 1991). Rehabilitation of A_p horizons (with massive to angular blocky structure) to "A_h horizons" with crumb and granular structure similar to no-till and/or minimum-till systems could be measured using trace elements and/or molecules following microscopic and submicroscopic techniques employed by Tomlin et al. (1993).

Acknowledgments

The authors acknowledge critical reading of the manuscript by Dr. C. Fox and the technical assistance of J. Miller, J. Fonseca, B. Vandenbygaart, G. Lambert, and library assistance of D. Drew.

References

Abbott, I. and C.A. Parker. 1981. Interactions between earthworms and their soil environment. *Soil Biol. Biochem.* 13:191–197.

Altemuller, H.J. 1974. Mikroskopie der boeden mit hilfe von duennschliffen (Soil microscopy with thin sections). *Handbuch der mikroskopie in der technik*. Band IV, Teil 2. Umschau Verlag, Frankfurt am Main. pp. 309–367.

Altemuller, H.-J., and M. Joschko. 1992. Fluorescent staining of earthworm casts for thin section microscopy. *Soil Biol. Biochem*. 24:1577–1582.

Altemuller, H.J. and B. van Vliet-Lanoe. 1990. Soil thin section fluorescence microscopy. p. 565–579. In: Douglas, L.A. (ed.) *Soil micromorphology*. Elsevier Science Publishers, Amsterdam.

Andreini, M.S., and R.S. Steenhuis. 1990. Preferential paths of flow under conventional and conservation tillage. *Geoderma* 46:85–102.

Baker, J.L. 1987. Hydrologic effects of conservation tillage and their importance relative to water quality. p. 113–124. In: Logan, T.J. et al. (eds.) *Effects of conservation tillage on groundwater quality: nitrates and pesticides*. Lewis Publications, Chelsea, MI.

Bal, L. 1973. *Micromorphological analysis of soils. Lower levels in the organization of organic soil materials*. Soil Survey Papers No.6. Soil Survey Institute, Wageningen, The Netherlands.

Bal, L. 1982. *Zoological ripening of soils*. Ctr. Agric. Publ. Doc., Report 850. Wageningen, The Netherlands.

Barley, K.P. 1959a. The influence of earthworms on soil fertility. II. Consumption of soil and organic matter by the earthworm *Allolobophora caliginosa*. *Aust. J. Agric. Res*. 10:179–185.

Barley, K.P. 1959b. Earthworms and soil fertility. IV. The influence of earthworms on the physical properties of a red-brown earth. *Aust. J. Agric. Res*. 10:371–376.

Barois, I., G. Villemin, P. Lavelle, and F. Toutain. 1993. Transformation of the soil structure through *Pontoscolex corethrurus* (Oligochaeta) intestinal tract. *Geoderma* 56:57–66.

Beare, M.H. and R.R. Bruce. 1993. A comparison of methods for measuring water-stable aggregates: Implications for determining environmental effects on soil structure. *Geoderma* 56:87–104.

Berry, E.C. and D.L. Karlen. 1993. Comparison of alternative farming systems. II. Earthworm population density and species diversity. *Am. J. Alt. Agric*. 8:21–26.

Beven, K. and P. Germann. 1982. Macropores and water flow in soils. *Water Resour. Res*. 19:1311–1325.

Bouché, M.B. 1971. Relations entre les structures spatiales et fonctionelles des écosystèmes, illustrées par le role pédobiologique des vers de terre. p. 189–209. In: Delamere Deboutteville, C. (ed.) *La Vie dans les Sols*. Gauthiers Villars, Paris.

Bouché, M.B. 1977. Stratégies lombriciennes. In: Lohm, U. and T. Persson (eds.) *Soil organisms as components of ecosystems*. Ecological Bulletin (Stockholm) 25:122–132.

Bouma, J. 1991. Influence of soil macroporosity on environmental quality. *Adv. Agron*. 46:1–37.

Bouma, J., C.F.M. Belmans, and L.W. Dekker. 1982. Water infiltration and redistribution an a silt loam subsoil with vertical worm channels. *Soil Sci. Soc. Am. J*. 46:917–921.

Bouma, J., C. Belmans, L.W. Dekker, and W.J.M. Jeurissen. 1983. Assessing the suitability of soils with macropores for subsurface liquid waste disposal. *J. Environ. Qual*. 12:305–311.

Bouma, J., L.W. Dekker, and C.J. Muilwijk. 1981. A field method for measuring short-circuiting in clay soils. *J. Hydrol*. 52:347–354.

Brusseau, N.L. and P.S.C. Rao. 1990. Modeling solute transport in structured soils: A review. *Geoderma* 46:169–192.

Bullock, P., N. Fedoroff, A. Jongerius, G. Stoops, T. Tursina, and U. Babel. 1985. *Handbook for soil thin section description*. Waine Research Publications. Albrighton, Wolverhampton, England. 152 p.

Coltman, K.M., L.C. Brown, N.R. Fausey, A.D. Ward, and T.J. Logan. 1991. Soil columns for solute transport studies: A review. *Am. Soc. Agric. Eng. Paper No.* 91-2150, 35 pp.

Darwin, C. 1881. *The formation of vegetable mould through the action of worms, with observations on their habits*. Murray, London.

Dawson, R.C. 1947. Earthworm microbiology and the formation of water-stable soil aggregates. *Soil Sci. Soc. Am. Proc.* 12:512–516.

De St. Remy, E.A. and T.B. Daynard. 1982. Effects of tillage methods on earthworm populations in monoculture corn. *Can. J. Soil Sci.* 62:699–703.

Douglas, J.T., M.J. Gross, and D. Hill. 1980. Measurements of pore characteristics in a clay soil under ploughing and direct drilling, including the use of a radioactive tracer (^{144}Ce) technique. *Soil Tillage Res.* 1:11–18.

Dutt, A.K. 1948. Earthworms and soil aggregation. *J. Am. Soc. Agron.* 40:407–410.

Edwards, A.P. and J.M. Bremner. 1967. Microaggregates in soils. *J. Soil Sci.* 18:64–73.

Edwards, C.A. 1980. Interactions between agricultural practice and earthworms. p. 3–12. In: Dindal, D.L. (ed.) *Soil biology as related to land use practices*. EPA, Washington, D.C.

Edwards, C.A. and G.W. Heath. 1975. Studies in leaf litter breakdown. III. The influence of leaf age. *Pedobiologia* 15:348–354.

Edwards, C.A. and J.R. Lofty. 1977. *Biology of earthworms*. 2nd ed. Chapman and Hall, London.

Edwards, C.A. and J.R. Lofty. 1982. The effect of direct drilling and minimal cultivation on earthworm populations. *J. Appl. Ecol.* 19:723–734.

Edwards, W.M., L.D. Norton, and C.E. Redmond. 1988a. Characterizing macropores that affect infiltration into nontilled soils. *Soil Sci. Soc. Am. J.* 43:851–856.

Edwards, W.M., M.J. Shipitalo, W.A. Dick, and L.B. Owens. 1992. Rainfall intensity affects transport of water and chemicals through macropores in no-till soil. *Soil Sci. Soc. Am. J.* 56:52–58.

Edwards, W.M., M.J. Shipitalo, and L.D. Norton. 1988b. Contribution of macroporosity to infiltration into a continuous corn no-tilled watershed: Implications for contaminant movement. *J. Contam. Hydrol.* 3:193–205.

Edwards, W.M., M.J. Shipitalo, and L.B. Owens. 1993a. Gas, water and solute transport in soils containing macropores: A review of methodology. *Geoderma* 57:31–49.

Edwards, W.M., M.J. Shipitalo, L.B. Owens, and W.A. Dick. 1993b. Factors affecting preferential flow of water and atrazine through earthworm burrows under continuous no-till corn. *J. Environ. Qual.* 22:453–457.

Edwards, W.M., M.J. Shipitalo, L.B. Owens, and L.D. Norton. 1989. Water and nitrate movement in earthworm burrows within long-term no-till corn fields. *J. Soil Water Conserv.* 44:240–243.

Ehlers, W. 1975. Observations on earthworm channels and infiltration on tilled and untilled loess soil. *Soil Sci.* 119:242–249.

Evans, A.C. 1948. Studies on the relationships between earthworms and soil fertility. II. Some effects of earthworms on soil structure. *Ann. Appl. Biol.* 35:1–13.

Everts, C.J., R.S. Kanwar, E.C. Alexander Jr., and S.C. Alexander. 1989. Comparison of tracer mobilities under laboratory and field conditions. *J. Environ. Qual.* 18:491–498.

Fitzpatrick, E.A. 1986. *An introduction to soil science.* 2nd ed. Longman, England.

Gantzer, C.J. and G.R. Blake. 1978. Physical characteristics of Le Sueur clay loam following no-till and conventional tillage. *Agron. J.* 70:853–857.

Germann, P.F., W.M. Edwards, and L.B. Owens. 1984. Profiles of bromide and increased soil moisture after infiltration into soils with macropores. *Soil Sci. Soc. Am. J.* 48:237–244.

Guild, W.J. McL. 1955. Earthworms and soil structure. p. 83–98. In: Kevan, D.K. McE. (ed.) *Soil zoology.* Butterworths, London.

Hamilton, W.E. and D.L. Dindal. 1989. Impact of landspread sewage sludge and earthworm introduction on established earthworms and soil structure. *Biol. Fertil. Soils* 8:160–165.

Hartenstein, F., E. Hartenstein, and R. Hartenstein. 1981. Gut load and transit time in the earthworm *Eisenia foetida. Pedobiologia* 22:5–20.

Hassink, J., L.A. Bouwman, K.B. Zwart, J. Bloem, and L. Brussaard. 1993. Relationships between soil texture, physical protection of organic matter and soil biota and C and N mineralization in grassland soils. *Geoderma* 57:105–128.

Heine, O. and O. Larink. 1993. Food and cast analyses as a parameter of turn-over of materials by earthworms (*Lumbricus terrestris* L.). *Pedobiologia* 37:245–256.

Hendriksen, N.B. 1990. Leaf litter selection by detritivore and geophagous earthworms. *Biol. Fertil. Soils* 10:17–21.

Hoeksema, K.J. and A. Jongerius. 1959. On the influence of earthworms on the soil structure in mulched orchards. *Proc. Int. Symp. Soil Struct. Ghent* 1958, pp. 188–194.

Hopp, H. and H. Hopkins. 1946. Earthworms as a factor in the formation of water-stable aggregates. *J. Soil Water Cons.* 1:11–13.

James, S.W. 1991. Soil, nitrogen, phosphorus, and organic matter processing by earthworms in a tallgrass prairie. *Ecology* 76:2101–2109.

Jandel Scientific. 1993. Mocha® Image Analysis Software for Windows. San Rafael, CA.

Joannes, H. and A. Kretzschmar. 1983. Model of gut transit in earthworms *Nicodrilus* spp. Rev. *Ecol. Biol. Soil* 20:349–366.

Jones, L.A., E.M. Rutledge, H.D. Scott, D.C. Wolf, and B.J. Teppen. 1993. Effects of two earthworm species on movement of septic tank effluent through soil columns. *J. Environ. Qual.* 22:52–57.

Joschko, M., W. Söchtig, and L. Larink. 1992. Functional relationship between earthworm burrows and soil water movement in column experiments. *Soil Biol. Biochem.* 24:1545–1547.

Kladivko, E.J., A.D. Mackay, and J.M Bradford. 1986. Earthworms as a factor in the reduction of soil crusting. *Soil Sci. Soc Am. J.* 50:191–196.

Knight, D., P.W. Elliott, J.M. Anderson, and D. Scholefield. 1992. The role of earthworms in managed, permanent pastures in Devon, England. *Soil Biol. Biochem.* 24:1511–1517.

Knollenberg, W.G., R.W. Merritt, and D.L. Lawson. 1985. Consumption of leaf litter by *Lumbricus terrestris* (Oligochaeta) on a Michigan woodland floodplain. *Am. Midl. Nat.* 113:1–6.

Kretzschmar, A. 1987. Soil partitioning effect of an earthworm burrow system. *Biol. Fertil. Soils.* 3:121–124.

Lal, R. 1974. No-tillage effects on soil properties and maize (*Zea mays* L.) production in western Nigeria. *Plant Soil* 40:237–244.

Lal, R. and O.O. Akinremi. 1983. Physical properties of earthworm casts and surface soil as influenced by management. *Soil Sci.* 135:114–122.

Lauren, J.G., R.J. Wagenet, J. Bouma, and J.H.M. Woosten. 1988. Variability of saturated hydraulic conductivity in a Glossaquic Hapludalf with macropores. *Soil Sci.* 145:20–28.

Lavelle, P., R. Schaefer, and Z. Zaidi. 1989. Soil ingestion and growth in *Millsonia anomala*, a tropical earthworm, as influenced by the quality of the organic matter ingested. *Pedobiologia* 33:379–388.

Lee, K.E. 1985. *Earthworms, their ecology and relationships with soils and land use.* Academic Press, New York.

Lee, K.E. and R.C. Foster. 1991. Soil fauna and soil structure. *Aust. J. Soil Res.* 29:745–775.

Low, A.J. 1972. The effect of cultivation on the structure and other physical characteristics of grassland and arable soils (1945–1970). *J. Soil Sci.* 23:363–380.

Mackay, A.D. and E.J. Kladivko. 1985. Earthworms and rate of breakdown of soybean and maize residues in soil. *Soil Biol. Biochem.* 17:851–857.

Madge, D.S. 1969. Field and laboratory studies on the activities of two species of tropical earthworms. *Pedobiologia* 9:188–214.

Marinissen, J.C.Y. and A.R. Dexter. 1990. Mechanisms of stabilization of earthworm casts and artificial casts. *Biol. Fertil. Soils* 9:163–167.

Martin, N.A. 1982. The interaction between organic matter in soil and the burrowing activity of three species of earthworms (Oligochaeta: Lumbricidae). *Pedobiologia* 24:185–190.

McCabe, D.C., R. Protz, and A.D. Tomlin. 1991. Faunal effects on the distribution of gamma emitting radionuclides in four forested soils. *Water, Air and Soil Pollution.* 57–58:521–532.

McKenzie, B.M. and A.R. Dexter. 1987. Physical properties of casts of the earthworm *Aporrectodea rosea*. *Biol. Fertil. Soils* 5:152–157.

McKenzie, B.M. and A.R. Dexter. 1993. Size and orientation of burrows made by the earthworms *Aporrectodea rosea* and *A. caliginosa*. *Geoderma.* 56:233–241.

Nielsen, D.R., M.Th. Van Genuchten, and J.W. Bigger. 1986. Water flow and solute transport processes in the unsaturated zone. *Water Resour. Res.* 22:89–109.

Nielson, R.L. 1951. Effect of soil minerals on earthworms. *N. Z. J. Agric.* 83:433–435.

Oades, J.M. 1993. The role of biology in the formation, stabilization and degradation of soil structure. *Geoderma* 56: 377–400.

Oades, J.M. and A.G. Waters. 1991. Aggregate hierarchy in soils. *Aust. J. Soil Res.* 29:815–828.

Parle, J.N. 1963a. Microorganisms in the intestines of earthworms. *J. Gen. Micro.* 31:1–11.

Parle, J.N. 1963b. A microbiological study of earthworm casts. *J. Gen. Micro.* 31:13–22.

Pedersen, J.C. and N.B. Hendriksen. 1993. Effect of passage through the intestinal tract of detritivore earthworms (*Lumbricus* spp.) on the number of selected Gram-negative and total bacteria. *Biol. Fertil. Soils.* 16: 227–232.

Petruk, W. 1989. *Short course on image analysis applied to mineral and earth sciences.* Mineralogical Association of Canada, Ottawa.

Raw, F. 1962. Studies of earthworm populations in orchards. I. Leaf burial in apple orchards. *Ann. Appl. Biol.* 50:389–404.

Reid, J.B. and M.J. Goss. 1981. Effect of living roots of different plant species on the aggregate stability of two arable soils. *J. Soil Sci.* 32:521–541.

Reynolds, J.W. 1977. *The earthworms (Lumbricidae and Sparganophilidae) of Ontario.* Life Sciences Miscellaneous Publications, Royal Ontario Museum, Toronto.

Satchell, J.E. 1967. Lumbricidae. p. 259–322. In: Burges, A. and F. Raw (eds.) *Soil biology.* Academic Press, London.

Sharpley, A.N., J.K. Syers, and J.A. Springett. 1979. Effect of surface-casting earthworms on the transport of phosphorous and nitrogen in surface runoff from pasture. *Soil Biol. Biochem.* 11:459–462.

Shaw, C. and S. Pawluk. 1986. The development of soil structure by *Octolasion tyrtaeum, Aporrectodea turgida* and *Lumbricus terrestris* in parent materials belonging to different textural classes. *Pedobiologia* 29:327–339.

Shipitalo, M.J., W.M. Edwards, W.A. Dick, and L.B. Owens. 1990. Initial storm effects on macropore transport of surface-applied chemicals in no-till soil. *Soil Sci. Soc. Am. J.* 54: 1530–1536.

Shipitalo, M.J., W.M. Edwards, and C.E. Redmond. 1994. Comparison of water movement and quality in earthworm burrows and pan lysimeters. *J. Environ. Qual.* 23:1345–1351.

Shipitalo, M.J. and R. Protz. 1987. Comparison of morphology and porosity of a soil under conventional and zero tillage. *Can. J. Soil Sci.* 67: 445–456.

Shipitalo, M.J. and R. Protz. 1988. Factors influencing the dispersibility of clay in worm casts. *Soil Sci. Soc. Am. J.* 52:764–769.

Shipitalo, M.J. and R. Protz. 1989. Chemistry and micromorphology of aggregation in earthworm casts. *Geoderma* 45:357–374.

Shipitalo, M.J., R. Protz, and A.D. Tomlin. 1988. Effect of diet on the feeding and casting activity of *Lumbricus terrestris* and *L. rubellus* in laboratory culture. *Soil Biol. Biochem.* 20:233–237.

Skjemstad, J.O., L.J. Janik, M.J. Head, and S.G. McClure. 1993. High energy ultraviolet photo-oxidation; a novel technique for studying physically protected organic matter in clay-and silt-sized aggregates. *J. Soil Sci.* 44:485–499.

Smettem, K.R.J. 1992. The relation of earthworms to soil hydraulic properties. *Soil Biol. Biochem.* 24:1539–1543.

Smettem, K.R.J. and N. Collis-George. 1985a. Statistical characterization of soil biopores using a soil peel method. *Geoderma* 36:27–36.

Smettem, K.R.J. and N. Collis-George. 1985b. The influence of cylindrical macropores on steady-state infiltration in a soil under pasture. *J. Hydrol.* 52:107–114.

Smith, M.S., G.W. Thomas, R.E. White, and D. Ritonga. 1985. Transport of *Escherichia coli* through intact and disturbed soil columns. *J. Environ. Qual.* 14:87–91.

Springett, J.A., R.A.J. Gray, and J.B. Reid. 1992. Effect of introducing earthworms into horticultural land previously denuded of earthworms. *Soil Biol. Biochem.* 24:1615–1622.

Springett, J.A. and J.K. Syers. 1984. Effect of pH and calcium content of soil on earthworm cast production in the laboratory. *Soil Biol. Biochem.* 16:185–189.

Stehouwer, R.C., W.A. Dick, and S.J. Traina. 1993. Characteristics of earthworm burrow lining affecting atrazine sorption. *J. Envir. Qual.* 22:181–185.

Stehouwer, R.C., W.A. Dick, and S.J. Traina. 1994. Sorption and retention of herbicides in vertically oriented earthworm and artificial burrows. *J. Environ. Qual.* 23:286–292.

Stephens, P.M., C.W. Davoren, M.H. Ryder, and B.M. Doube. 1993. Influence of the lumbricid earthworm *Aporrectodea trapezoides* on the colonization of wheat roots by *Pseudomonas corrugata* strain 2140R in soil. *Soil Biol. Biochem.* 25:1719–1724.

Swaby, R.J. 1950. The influence of earthworms on soil aggregation. *J. Soil Sci.* 1:195–197.

Teotia, S.P., F.L. Duley, and T.M. McCalla. 1950. Effect of stubble mulching on number and activity of earthworms. *Neb. Agr. Exp. Stn. Res. Bull.* 165:1–20.

Thomson, A.J. and D.M. Davies. 1974. Production of surface casts by the earthworm *Eisenia rosea*. *Can. J. Zool.* 52:659.

Tisdall, J.M. and J.M. Oades. 1982. Organic matter and water-stable aggregates in soils. *J. Soil Sci.* 33:141–163.

Tomlin, A.D., R. Protz, R.R. Martin, D.C. McCabe, and R.J. Lagace. 1993. Relationships amongst organic matter content, heavy metal concentrations, earthworm activity, and soil microfabric on a sewage sludge disposal site. *Geoderma* 57:89–103.

Trojan, M.D. and D.R. Linden. 1992. Microrelief and rainfall effects on water and solute movement in earthworm burrows. *Soil Sci. Soc. Am. J.* 56:727–733.

Tyler, D.D. and G.W. Thomas. 1981. Chloride movement in undisturbed soil columns. *Soil Sci. Soc. Am. J.* 45:459–461.

Utomo, W.H. and A.R. Dexter. 1982. Changes in soil aggregate water stability induced by wetting and drying cycles in non-saturated soil. *J. Soil Sci.* 33:623–637.

Van Rhee, J.A. 1963. Earthworm activities and the breakdown of organic matter in agricultural soils. p. 55–59. In: Doeksen, J. and J. van der Drift (eds.) *Soil organisms*. North Holland, Amsterdam.

Wagenet, R.J. 1986. Water and solute flux. p 1055–1088. In: Klute, A. (ed.) *Methods of soil analysis, Part I*. 2nd ed. Agron. Monogr. 9, ASA and SSSA, Madison, WI.

Watanabe, H. and S. Ruaysoongnern. 1984. Cast production by the Megascolecid earthworm *Pheretima* sp. in northeastern Thailand. *Pedobiologia* 26:37–44.

Wershaw, R.L. 1992. Membrane-micelle model for humus in soils and sediments and its relation to humification. U.S. Geological Survey Open-File Report 91-513. Denver, CO. 64 pp.

West, L.T., P.F. Hendrix, and R.R. Bruce. 1991. Micromorphic observation of soil alteration by earthworms. *Agric. Ecosystems Environ.* 34:363–370.

Zachmann, J.E. and D.R. Linden. 1989. Earthworm effects on corn residue breakdown and infiltration. *Soil Sci. Soc. Am. J.* 53:1846–1849.

Zachmann, J.E., D.R. Linden, and C.E. Clapp. 1987. Macroporous infiltration and redistribution as affected by earthworms, tillage, and residue. *Soil Sci. Soc. Am. J.* 51:1580–1586.

Earthworms in Agroecosystems

Clive A. Edwards, Patrick J. Bohlen,
Dennis R. Linden, and Scott Subler

1-56670-053-1/95/$0.00+$.50
©1995 by CRC Press, Inc.

I.   Introduction

Earthworms are probably the most important soil-inhabiting invertebrates in terms of their influence on soil formation and maintenance of soil fertility in agroecosystems. Although they are not numerically dominant, their large size and habit of burrowing through and turning over large quantities of soil make them major contributors to overall soil structure, aeration, and drainage. Their consumption of many forms of organic matter, and their interactions with the soil microbial community, increase the availability of nutrients, and play a major role in the recycling of carbon and nitrogen in soils (Darwin 1881, Edwards and Lofty 1977, Lee 1985, Edwards and Bohlen 1995; see also Blair et al. and Tomlin et al., this volume).

The importance of earthworms to soil fertility has been emphasized in recent years by changes that are occurring in the management of both tropical and temperate agroecosystems. The intensity of tillage has decreased progressively in the last 30 to 40 years, ranging through practices such as shallow or chisel-plowing, to a complete absence of tillage (no-till) (Allmaras et al. 1994). Earthworms thrive in many of these situations, and are responsible for incorporating large amounts of crop residues into the soil, and increasing water infiltration and drainage significantly, thereby minimizing surface erosion (Edwards and Lofty 1982b). Earthworm populations also increase in size as agriculture moves away from its dependence on large amounts of inorganic fertilizers and pesticides, substituting for them with more rotations and biological and cultural sources of fertility and pest control. It seems clear that the impacts of earthworms on crop productivity and soil properties are bound to increase in importance as these trends continue.

Although some excellent contributions to our understanding of the effects of earthworms on agroecosystems have come from North America (e.g., Hopp and Slater 1948, 1949, Mackay and Kladivko 1985, Parmelee and Crossley 1988, Edwards et al. 1989), studies of earthworms in agroecosystems on this continent are relatively few in number. Clearly, there is a need for much more research on earthworms in the diverse agroecosystems of North America. In this paper, we provide an overview of the biology and ecology of earthworms in agroeco-systems, stressing the importance of earthworms to soil fertility and their relationships with management practices. We cite examples from North America, where possible, and recommend key areas for future research.

II.   Biology and Ecology of Earthworms in Agroecosystems

A.   Species Diversity and Associations

There are about 3000 species of earthworms belonging to the class Oligochaeta distributed throughout the world, although there is considerable controversy on their systematics and many are aquatic in habit. They are common in most parts of the world, except in areas with extreme climates, such as deserts, previously glaciated land, and areas under constant snow and ice. Many of the earthworm species that are important to agriculture in North America, particularly those belonging to the family Lumbricidae, are not endemic but have been introduced from Europe and are distributed widely ("peregrine") (see Reynolds, this volume). These earthworms tend to become dominant in ecosystems to which they are introduced, either because of the lack of endemic species in previously glaciated or disturbed areas, or because they are strong competitors for available resources (see Kalisz and Wood, this volume). This situation applies to many North American agroecosystems, although the distribution of neither the introduced nor the endemic earthworm fauna of North America has been well studied, and needs further research, particularly in the western United States.

The diversity of species of earthworms varies greatly and there tend to be species associations in different soil types and habitats. The associations of species of lumbricids in temperate agricultural soils (Edwards and Lofty 1982b) are typically less diverse than those of earthworms from other families in warmer latitudes (Lavelle 1992). However, even in the more diverse communities, there are commonly only 4 to 6 species, and rarely more than 10, at a given location. Competition between earthworm species and pressure from predators are probably less important in regulating earthworm populations than variability in environmental conditions (Lee 1985).

B.   Life Cycles

The life cycles of different earthworm species differ considerably and we still lack detailed knowledge of the biology and ecology of many species. Earthworms are hermaphrodites, and mate either on or in the soil. During mating, both individuals become fertilized and each then produces a cocoon which is formed from the clitellum. The cocoon is passed over the earthworm's head and deposited in the soil, where it hardens into a characteristic shape that is distinctive for each species. Each cocoon may contain a single egg or several eggs, depending on the species involved. Cocoons can be produced throughout the year but, in temperate climates, most cocoons are produced in spring (Edwards and Lofty 1977). The number of cocoons produced annually by a single earthworm ranges from about twenty to several hundreds, depending on the species involved, and the suitability of environmental conditions. Cocoons resist desiccation and may be viable in soil for several years, even under adverse soil conditions, hatching when soil conditions become favorable. Under favorable soil conditions, the time of incubation of eggs in the cocoons ranges from 8 to 20 weeks. Newly-hatched worms take from 10 weeks to about 1 year to reach sexual maturity. Although earthworms can live for up to 10 years, few probably survive more than 1 to 2 years under field conditions. Some species can reproduce parthenogenetically without mating.

C.   Population Size and Variability

The size of earthworm populations ranges from only a few to more than 1000 m^{-2}, and depends on a wide range of factors (Edwards and Lofty 1977, Lee 1985). Typical populations seldom exceed 100 to 200 m^{-2} in cultivated soil or 400 to 500 m^{-2} in grassland (Edwards 1983). Large populations, of up to 500 m^{-2}, are sometimes found in orchards, where the availability of organic matter is seldom limiting. Earthworm populations in soils under some form of conservation tillage or no-till are usually much higher than in plowed or intensively cultivated soils (Edwards and Lofty 1982b).

The factors that influence the size of earthworm populations in agroecosystems include: organic matter status, soil type, pH, moisture-holding capacity, rainfall, temperature, cultivations, crop residues, and cropping patterns. Among these factors, the availability of organic matter is probably the most important (Edwards and Lofty 1982b, Lee 1985). For instance, Hendrix et al. (1992) reported a significant positive correlation between earthworm numbers and soil organic matter content across a variety of ecosystem types on the Georgia piedmont, in the southeastern United States.

It is widely accepted that most cultivated soils in temperate regions support earthworm populations. However, the actual number of studies concerning the distribution and abundance of earthworms in North American agricultural soils is surprisingly small, especially considering the importance of earthworms to soil fertility in many regions. Earthworm populations and species compositions often differ greatly among agricultural sites and we are only beginning to understand many of the factors that influence species distributions (Bohlen et al. 1994, Hendrix et al. 1992).

D.   Population Dynamics

Earthworm population dynamics are relatively complex, but they depend principally upon the suitability of soil moisture and temperature for development and activity (Edwards and Lofty 1977). In periods of intense precipitation, earthworms often emerge from their burrows in large numbers, and migrate over the soil surface. The reasons for these mass migrations are unknown but may be related to lack of oxygen or other chemical conditions that exist in soils completely saturated by rainfall, or to other behavioral cues that lead to dispersal. Individuals of *Lumbricus terrestris* have been observed to migrate up to 20 m in a single night (Mather and Christensen 1988).

In temperate regions, earthworms are most active in spring and fall. During winter, they retreat to the deeper soil layers to escape adverse temperature conditions, though they can become quite active again during cool, wet periods when the ground is not frozen. In spring, the peak production of cocoons usually occurs. Earthworms can survive summer drought either as cocoons or by burrowing deep into the soil. Some species construct cells lined with mucus fairly deep in the soil, in which they aestivate in a coiled position during hot, dry periods (Edwards and Lofty 1977). They remain in these cells until the moisture and temperature conditions are favorable for renewed activity. In the humid tropics, earthworm activity is also linked with availability of moisture and suitability of temperature, and there is a strong peak of activity during hot wet seasons (Lavelle and Fragoso 1992).

III. Earthworms and Soil Fertility

Many species of earthworm have been implicated in the maintenance and improvement of soil fertility and crop productivity (Hopp and Slater 1949, Barley 1959a,b, Stockdill 1982, Edwards 1983, Syers and Springett 1984). Earthworm activity usually increases the availability of nutrients, accelerates mineralization of organic matter, and improves soil structure (Edwards and Lofty 1977, Lee 1985). Other benefits of earthworm activity include the suppression of certain pest or disease organisms and the enhancement of numbers of beneficial microorganisms in soil (Doube et al. 1994). Most investigations have reported increases in crop productivity in the presence of earthworms, or in response to earthworm inoculation, although this has not always been observed. Occasionally, earthworms can have undesirable consequences in agroecosystems, such as by increasing the gaseous losses of nitrogen due to denitrification (Svensson et al. 1986, Elliot et al. 1990, Knight et al. 1992), or by increasing leaching of nitrate, especially when nitrogenous fertilizers are applied to the soil surface (Knight et al. 1992). However, it is generally accepted that any adverse effects of earthworms on soil processes are greatly outweighed by the beneficial effects of earthworms.

Earthworms affect soil fertility by changing the physical, chemical, and biological properties of soils. The effects of earthworms may be a direct result of their feeding and burrowing activities, or an indirect result of their numerous interactions with soil microorganisms and dynamic soil processes. Some of the effects of earthworms on soil processes are seasonal and depend on the feeding behavior and life history characteristics of different species. Some influences of earthworms on soil processes are small-scale and ephemeral, whereas others are large-scale and persistent, and it is often very difficult to extrapolate from small, localized effects to the more important large-scale effects at the ecosystem level. The degree of influence of earthworms on soil fertility also depends on the soil type, climate, and the nutrient and organic matter status of the soil. In the following sections we will summarize what is known about the influence of earthworms on key soil processes and features as they relate to soil fertility and crop productivity. Other chapters in this volume provide more detailed accounts of the influences of earthworms on biogeochemistry (Blair et al., this volume) and soil structure (Tomlin et al., this volume).

A.   Soil Turnover

Earthworms can ingest and turn over substantial amounts of soil (Darwin 1881, Edwards and Lofty 1977, Lee 1985), but it is difficult to make accurate estimates of the total amounts of soil that they turn over annually at any given location. Furthermore, most estimates of soil turnover by earthworms are for grasslands and pastures, and we know much less about the amounts of soil turned over by earthworms in cultivated soils.

The rates of total cast production by several species have been determined, but usually only in pot studies. Barley (1959a) calculated the rate of cast production by *Aporrectodea caliginosa* to be 0.31 g g^{-1} fresh wt d^{-1} which he calculated to be equivalent to 3 to 4 kg m^{-2} yr^{-1}. Bolton and Phillipson (1976) calculated the cast egestion rate of *A. rosea* to be 1 to 2 g g^{-1} fresh weight of worm. The total amount of soil egested by an earthworm community can be as high as 50 kg dry soil m^{-2} yr^{-1} for temperate grasslands (Lavelle 1988). Annual production of surface casts by earthworms in temperate, European pastures, ranges from 0.7 to 25.7 kg m^{-2} (Lee 1985). Syers et al. (1979) estimated that the surface cast production in New Zealand pastures was from 2.5 to 3.3 kg m^{-2} yr^{-1}. Graff (1971) estimated that the production of surface casts by earthworms in a German pasture amounted to 25% of the soil in the A$_h$ horizon. These estimates of the total turnover of soil by earthworms are conservative, because they are based solely on surface cast production. Clearly, relatively large amounts of soil can be egested by earthworms, which must have a large impact on soil structure and the availability of nutrients in many agroecosystems.

B.   Decomposition and Redistribution of Organic Matter

Earthworms can have an enormous influence on the dynamics and decomposition of organic matter in agroecosystems. Mackay and Kladivko (1985) reported that *Lumbricus rubellus* increased the rates of disappearance of soybean residues from the soil surface by 165% and corn residues by 320%, in a 36-day pot experiment. Syers et al. (1979) estimated that a mixed community of *L. rubellus* and *A. caliginosa* removed over 80 kg of surface litter ha^{-1} day^{-1} in a New Zealand pasture, during the period of their highest seasonal activity, and a total of 6100 kg of litter ha^{-1} yr^{-1}. Parmelee et al. (1990) used carbofuran to reduce earthworm populations in no-till plots by more than 90% and reported that the amounts of fine, coarse, and total particulate organic matter increased by 43, 30, and 32%, respectively, 292 days after applying the carbofuran. This suggests that earthworms increased the mineralization of C from the pool of particulate organic matter in soil. Consequently, soils that have small natural populations of earthworms often develop a mor structure with a mat of undecomposed organic matter close to the soil surface, particularly in grassland and orchards (Kleinig 1966, Potter et al. 1990). Such mats are common on poor upland grasslands in temperate regions and in New Zealand in areas to which earthworms have not yet been introduced (Stockdill 1966).

The influence of earthworms on the rates of decomposition of organic matter is related mainly to the feeding behavior of different species. Some species, such as *Lumbricus terrestris*, an earthworm occurring commonly in temperate agroecosystems, feed mainly on surface plant litter. Individuals of *L. terrestris* live in permanent vertical burrows that may extend more than two meters deep into the soil, and incorporate surface plant residues into the soil, often forming piles of residue or "middens" around the mouth of their burrows. Feeding by *L. terrestris* can significantly increase the rate of disappearance of surface plant residues and their incorporation into the soil, as well as the rate at which nutrients are released from the residues into the soil (Binet and Trehen 1992). In orchards, populations of *L. terrestris* has been reported to consume all of the annual leaf fall of 2000 kg ha^{-1} yr^{-1} (Raw 1962). In no-till soils they are important because they can incorporate a significant amount of crop residues into the soil (Mackay and Kladivko 1985). The feeding activities of *L. terrestris* may even make it difficult to maintain the 30% cover of protective crop residues, which is required for cultivation practices to be classified as conservation tillage in the U.S. *L. rubellus* also feeds mainly on surface plant or crop residues, but unlike *L. terrestris*, burrows continuously through the upper horizons of soil rather than forming a permanent vertical burrow.

Other species of earthworm in North America (e.g., *Aporrectodea* spp., *Octolasion* spp., *Diplocardia* spp.), whose diets contain a much higher proportion of soil than the diet of the worms that feed primarily on surface residues, have different effects on organic matter dynamics than do the surface feeders (Piearce 1978). These soil-ingesting, or geophagous, species feed selectively on soil enriched in organic matter (Bolton and Phillipson 1976), although they may also feed to some extent on surface crop residues. They burrow continuously through the upper soil horizons, altering the association of organic matter with mineral particles and mixing it more evenly throughout the soil volume. Their casts may be egested either onto or below the soil surface, but most are probably egested below the soil surface.

Litter-feeding and geophagous earthworms usually occur together in most soils and they probably have a synergistic effect on the redistribution of organic matter throughout the soil profile. Shaw and Pawluk (1986b) reported that when *L. terrestris* and *O. cyaneum* were in soil microcosms together, they distributed the surface crop residues more evenly throughout the soil matrix, than when either species was present alone. This process may be important in no-till, minimum till, and perennially-cropped agroecosystems, in which surface organic matter could be incorporated more thoroughly throughout the soil profile by an earthworm community containing both geophagous and surface-litter-feeding earthworms, compared with a community containing only one of these groups.

There is some evidence that the organic matter in earthworm casts may be protected from decomposition. Earthworms can alter the association of organic materials and soil particles, consuming a greater proportion of finer soil particles than is present in the surrounding soil (Mulongoy and Bedoret 1989, Marinissen and Dexter 1990); this may lead to the stabilization of organic matter in casts. Lavelle and Martin (1992) showed that earthworm casts had lower respiration rates than reference soil in long-term incubations, suggesting that organic matter in earthworm casts may be protected from decomposition as the casts age.

C.   Nutrient Cycling and Availability

Significant amounts of nutrients can pass directly through the earthworm biomass in agroecosystems. The direct flux of nitrogen through earthworm biomass was estimated to be 63 kg N ha^{-1} yr^{-1} in a no-till agroecosystem in Georgia (Parmelee and Crossley 1988) and 20 to 42 kg N ha^{-1} during the autumn in three arable systems in Denmark (Christensen 1988). Nitrogen in earthworm tissues turns over rapidly (Blair et al., this volume) and nutrients in dead earthworm tissue are mineralized rapidly (Satchell 1967).

Earthworms also return N to the soil in the form of urine and mucoproteins, which are both readily assimilable forms of N. Lee (1983) estimated that an average population of lumbricids may produce around 18 to 50 kg N ha^{-1} yr^{-1} in their urine. Less is known about the amounts of mucus produced in earthworm casts or secreted from their body walls. Scheu (1991) estimated that *Octolasion lacteum* produced, in one day, an amount of mucus equivalent to about 0.2% of total carbon and nitrogen. The combined direct contributions of earthworm tissues, urine, and mucus secretions to the cycling of nutrients can be significant, but the main influences of earthworms on nutrient transformations may be due to indirect effects of their burrowing, feeding, and casting activities.

Plant nutrients usually occur in higher concentrations in fresh earthworm casts and around the lining of their burrows than in bulk soil (Parle 1963). Concentrations of nitrogen (Lavelle and Martin 1992), phosphorus (Lunt and Jacobson 1944, Mansell et al. 1981, Mackay et al. 1982, Tiwari et al. 1989), and potassium (Tiwari et al. 1989, Basker et al. 1992), in forms available to plants, tend to be higher in fresh earthworm casts than in bulk soil. The increase in available nutrients in fresh earthworm casts lasts for a relatively brief period of time (Lavelle and Martin 1992). Earthworms also usually increase the availability of nutrients in bulk soil, but this is dependent upon a variety of factors, especially the amounts and types of organic matter in the soil and the kind of nutrient amendments that are added.

Earthworms can increase the total uptake of nutrients by plants by increasing the amounts of available nutrients in soil, and by secreting compounds in their casts that act like plant hormones and stimulate plant growth (Tomati et al. 1988). Stephens et al. (1994) reported that addition of *A. rosea* to soils containing wheat seedlings increased the foliar concentrations of Ca, Cu, and Mn in the seedlings, and that the addition of *A. trapezoides* increased the foliar concentrations of aluminum, calcium, iron, and manganese. McColl et al. (1982) reported that activities of *A. caliginosa* significantly increased the total uptake of most major and trace elements (N, P, K, Ca, Cl, Mg, Zn) by ryegrass seedlings, but did not increase the concentrations of these elements in the seedlings. Holmes (1952) reported that earthworms increased the growth of white clover in soils that were high in total but low in available Mo. Molybdenum is an element essential for nitrogen fixation, and he conjectured that the beneficial effects of earthworms on clover growth were caused by an increase in the amounts of available Mo.

Plant nutrient demand in agroecosystems must be synchronized with periods of earthworm activity, if management systems are to take optimal advantage of the nutrients that are made available to plants by earthworm activity. In permanent pastures near Adelaide in Australia, Barley (1959b) reported that the period of the most intense surface-casting by earthworms corresponded with the most rapid growth period of the grasses in the pasture. Presumably, the nutrients in the casts were available for uptake by these grasses during their most intense growth period. In many temperate and subtropical agroecosystems in North America, the most intense periods of earthworm activity, in spring and autumn, are not necessarily the periods of the most active crop uptake of soil nutrients. This is true particularly for summer annual crops. Seasonal earthworm activity may be synchronized better with the nutrient uptake by pastures and other perennial crops than with that by annual crops which have a brief peak demand for available nutrients. By increasing the availability of nutrients to crops when there is little actual uptake, earthworm activity may increase the potential for leaching and runoff of the more mobile nutrients such as nitrate. However, this has less chance of occurring with crops planted in fall or early spring, which have nutrient demands that are synchronized much better with the most intense periods of earthworm activity. Interestingly, James (1991) suggests that native earthworms are better adapted than introduced lumbricids to high soil temperatures and therefore may have a greater impact on annual nutrient cycles in tallgrass prairie in Kansas.

Earthworms have different effects on nutrient availability in agroecosystems when nutrients from different sources are added to the soil. Data from some of our recent field and laboratory experiments in Ohio have shown that earthworms can greatly increase the levels of available nitrogen in systems that have been fertilized with inorganic fertilizers, but have much less of an effect on levels of available nitrogen in systems fertilized with animal or green manures (Bohlen and Edwards, in press). This may have consequences for the leaching of nitrates in groundwater or loss of nitrogen through denitrification. We noted that there were higher concentrations of nitrate in the deeper (30 to 60 cm) soil horizons in plots with increased earthworm populations than in plots with reduced or unmanipulated populations. This effect was much greater in inorganically-fertilized plots than in organically-fertilized plots.

Earthworms can also increase the loss of nitrogen by increasing the rates of denitrification and nitrate leaching in soil. Fresh earthworm casts usually have higher denitrification rates than surrounding soil (Svensson et al. 1986, Elliot et al. 1990). Syers et al. (1979) reported a potential for increased loss of N from a New Zealand pasture in the cooler winter months due to denitrification in earthworm casts. Knight et al. (1992) estimated that earthworm casts on the soil surface in English pastures could account for 12% of the total denitrification losses from an unfertilized pasture and 26% of the losses from a fertilized pasture. They also showed that earthworms tripled the amounts of nitrate in leachates from these pastures. Again, the degree to which earthworms increased the losses of nitrogen depended on the amounts and types of fertilizer added, with losses being greater when large amounts of inorganic fertilizer were added to the soil. It is likely that these losses could be reduced by careful management of fertilizer inputs.

D.   Soil Microbial Activity

Earthworms can influence soil microbial biomass and activity in ways that impact both nutrient availability and mineralization rates. Microbial populations and biomass in earthworm casts are usually much higher than those in the surrounding soil (Shaw and Pawluk 1986a, Scheu 1987, Tiwari et al. 1989, Lavelle and Martin 1992). As was the case with available nutrients, pulses of microbial activity may be relatively transient (Lavelle and Martin 1992). In the long term, earthworm casts may have a lower overall microbial activity and biomass than the surrounding soil (Lavelle and Martin 1992), as substrates for microbial growth become depleted or are protected in stabilized earthworm casts (Martin 1991). Wolters and Joergensen (1992) suggested that earthworms may alter the composition of the soil microbial biomass, resulting in a smaller total biomass with a higher metabolic activity per unit of biomass. Soil microorganisms are essential components of the earthworm diet (Edwards and Fletcher 1988) and earthworms may change the ratio of fungi to bacteria, by feeding preferentially on particular groups of microorganisms and stimulating the growth of other groups in earthworm casts.

Earthworms accelerate the turnover of microbial populations and can release some of the nutrients bound up in the microbial biomass and make them available for plant growth. Spain et al. (1992) reported that the uptake of ^{15}N by *Panicum maximum,* from ^{15}N-labelled soils, increased significantly in soil to which earthworms were added, compared with soil to which no worms were added. Some of our recent field and laboratory experiments have shown that earthworms can decrease microbial-biomass-N, and increase available N in soil. Such results indicate that earthworms can increase the overall rates of transfer of nutrients from the microbial biomass to plants.

IV.   Effects of Earthworms on Beneficial Microorganisms

Earthworms have been shown to have a significant influence on the dispersal of vesicular-arbuscular mycorrhizae (VAM) fungi, which form an important mutualistic association with plant roots that can improve plant growth. Reddell and Spain (1991) surveyed the casts of 13 earthworm species from 60 sites in Australia and reported that intact spores of VAM fungi occurred in all but one batch of casts. They also found VAM-infected root fragments in the casts. The diversity of mycorrhizal fungi in earthworm casts was similar to that in surrounding soil, but the numbers of fungal spores was greatest in the casts of *Pontoscolex corethrurus* and *Diplotrema heteropora*. Greenhouse experiments verified that VAM spores and some root fragments recovered from casts could initiate mycorrhizal infection in *Sorghum bicolor*.

Earthworms can also influence populations of *Rhizobium* bacteria, which fix nitrogen in nodules on the roots of leguminous plants, including some important crop plants. *Lumbricus rubellus* enhanced the translocation of *Bradyrhizobium japonicum* to greater depths in soil (Madsen and Alexander 1982). Rouelle (1983) reported that *Lumbricus terrestris* increased the spread of root nodules on soybean (*Medicago sativa*). Doube et al. (1994) demonstrated that several species of earthworm could increase the degree of dispersal of *Rhizobium* bacteria and the amounts of root nodulation in alfalfa plants. They mixed a strain of *R. meliloti* with a dung-soil mixture and placed the inoculated mixture on the soil surface. They reported that numbers of bacteria associated with alfalfa roots, at a 3 to 9 cm soil depth, increased dramatically in the presence of *A. trapezoides*. There were also three times as many root nodules in pots with worms than in pots with no worms. It remains to be demonstrated whether such interactions between earthworms and beneficial microorganisms can influence crop growth in the field.

V.   Influence of Earthworms on Soil Structure and Physical Properties

Two of the most important structural features of agricultural soils that are influenced by earthworms are soil aggregation and macroporosity; hence, most research on the relationships between earthworms and soil structure has focused on these two areas. Earthworm activity has often been associated with improved soil aggregation (Hopp and Hopkins 1946, Stewart et al. 1980, Blanchart et al. 1989) and increased macroporosity (Hoeksema and Jongerius 1959, Ehlers 1975, Lee and Foster 1992), although the connection between these improved structural features and crop productivity has rarely been established. These changes in the structural features of soil caused by earthworms have been associated with increased water infiltration (Zachman et al. 1987, Zachman and

Linden 1989) and water-holding capacity, improved soil tilth, and a reduction in soil crusting (Kladivko et al. 1986).

Interactions between earthworm activity and soil physical properties are complex, and the overall effects of earthworms depend on the specific characteristics of the soil and the behavior and activity of the earthworm species being considered. We will concentrate on summarizing the influence of earthworms on key dynamic soil processes in agroecosystems that are affected by soil aggregate stability and macroporosity. A comprehensive review of earthworm effects on soil structure is given in Tomlin et al., this volume.

A.   Soil Aggregate Stability

The most direct way that earthworms contribute to the stability of soil aggregates is through the production of casts (Oades 1993). Fresh earthworm casts are often highly-dispersed, nearly-saturated masses of soil which are unstable and susceptible to erosion. As earthworm casts age, various physical, chemical, and biological processes influence their stabilization. Their organic matter content (Shipitalo and Protz 1988), wet-dry cycles (Marinissen and Dexter 1990), and age all enhance the development of a more stable cast structure (Parle 1963, Marinissen and Dexter 1990). Fungal hyphae (Marinissen and Dexter 1990) and other microbial products (Swaby 1950) can also help to stabilize casts. Stabilized casts contribute significantly to the improved aggregation of soil, which has often been observed after the introduction of earthworms to soil.

The walls of earthworm burrows can also contribute to the stability of soil structure. Earthworms secrete copious amounts of epidermal mucus, which is rich in mucopolysaccharides and proteins that can bind soil particles together. The linings of permanent earthworm burrows, such as those formed by *L. terrestris*, consist of oriented clay particles surrounded by concentric rings of humic materials (Jeanson 1964) which can form a stable structure.

The role of earthworms in improving soil aggregation are influenced strongly by reduced tillage and the addition of organic matter. Heavy tillage disrupts the aggregation of soil and can cause it to become compacted and much less macroporous. Heavy tillage also has a negative effect on earthworm populations. Adequate levels of organic matter are essential for both good soil aggregation (Tisdall and Oades 1982) and the development of large populations of earthworms. Current agricultural practices often limit the amounts of organic matter added to soil, which restricts the development of earthworm populations and optimal soil structure.

Earthworms are not essential for soils to be able to attain a good structure, and many soils are well-structured without earthworms. However, where earthworms are present, they usually have a predominant influence on soil aggregation. Much more research needs to be directed towards understanding the ways in which earthworm can modify soil structure, and how these modifications relate to long-term soil fertility, crop growth, and productivity.

B.   Soil Porosity

Earthworms have often been correlated positively with increased soil porosity (Teotia et al. 1950, Hoeksema and Jongerius 1959, Satchell 1967, Ehlers 1975). Hoeksema and Jongerius (1959), working in Dutch orchards, estimated that earthworms could increase the pore space in soil by 75 to 100% and Satchell (1967) estimated that earthworm burrows could account for up to two-thirds of the air-filled pores in soil. Hopp (1973) reported a positive correlation between earthworm populations and the proportion of large pores in crop and pastures soils, in the midwestern United States, in the spring. More recent studies in Ohio showed a significant increase in large macropores at a 10 cm depth, in soils of field plots in which earthworm population levels had been artificially increased (R.W. Parmelee, unpublished data). Other studies have also noted a significant positive correlation between the numbers and size of macropores and earthworm populations in field plots (Fuchs, unpublished data). One effect of earthworms, related to soil porosity, is to improve conditions in soil for increased crop growth by increasing the soil's water-holding capacity (van Rhee 1969, Stockdill 1966), water infiltration (Lee 1985), and soil aeration (Kretzschmar 1978, 1982).

The increased soil porosity and water infiltration produced by earthworms decreases the quantity of surface runoff. Syers and Springett (1984) used carbaryl to eliminate earthworms from a New Zealand pasture and reported that this resulted in a twofold increase in the volume of runoff and a threefold decrease in water infiltration rates. The total loss of N and P in surface runoff increased significantly after earthworms were eliminated (Sharpley et al. 1979).

Earthworm burrows can also influence the quality of infiltrating water and the potential leaching of nutrients and chemicals from agricultural land, because of their dramatic influence on water infiltration and preferential solute flow. The rate of flow of water through earthworm burrows during rainfall events is influenced by a variety of factors, the most important of which are the soil moisture content and rainfall intensity (Edwards et al. 1992, Trojan and Linden 1992). Fertilizers and pesticides applied to the soil surface may be transported well below the rooting zone through earthworm burrows. The likelihood that surface-applied chemicals will move down through earthworm burrows is greatest immediately after application, and diminishes progressively after chemicals have been incorporated into the soil by rainfall. Nitrate (and ammonium) concentrations in water flowing down burrows vary considerably and can be high at any time of year, not only in the period immediately following the application of inorganic fertilizer (Shipitalo et al. 1994).

VI. Effects of Earthworms on Crop Productivity

A.   Root Growth

There have been very few detailed studies which have investigated the influence
of earthworm activity on the growth of plant roots. Edwards and Lofty (1978)
removed intact soil profiles from a no-till field and inoculated them with either
a mixture of *L. terrestris* and *A. longa* or a mixture of *A. caliginosa* and *A.
chlorotica*. They compared the growth of barley roots in these soil profiles with
that in soil profiles with no earthworms. The root growth of the barley seedlings
in the soil profile with earthworms was correlated strongly with the zones of
activities of the earthworms, for all species of earthworm introduced to the
profiles. van Rhee (1977) monitored orchards that had been inoculated with
earthworms (mainly *A. caliginosa* and *L. terrestris*) and reported that the number
of small roots (<0.5 mm) increased at inoculated sites and decreased at
uninoculated sites, over an eight year period. The number of larger roots (>0.5
mm) was unaffected by earthworm inoculation. J. Pitkanen (pers. comm.)
reported that up to 60% of the earthworm burrows in the root zone of cereal
fields in Finland contained crop roots, which also suggests an important role of
earthworm biopores in influencing root growth.

Roots are known to grow preferentially along burrows and cracks in soil and
have been shown to proliferate around the casts of some earthworm species
(Springett and Syers 1979). Well-aerated, nutrient-rich earthworm burrows, can
provide a favorable environment for root growth (Edwards and Lofty 1980).
Such burrows may have been responsible for the development of grass root
systems to greater depths in soil and for large increases in yields following the
inoculation of earthworms into New Zealand pastures (Stockdill 1966, Lee
1985). Relatively few direct links have been made between the location of
earthworm activity and root growth (Logsdon and Linden 1992) and there is a
need for much more research on this topic.

B.   Inoculation of Earthworms into Agricultural Soils

Research in many parts of the world has shown that inoculation of earthworms
into soils with low populations can produce significant increases in yields of
crops. Many workers have investigated the effects of inoculating worms into
soils in pot experiments (Wollny 1890, Chadwick and Bradley 1948, Baluev
1950, Joshi and Kelkar 1952, Nielson 1953, Spain et al. 1992) and reported
considerable increases in root and shoot growth. Aldag and Graff (1974)
compared the growth of oat seedlings in pots with 800 g of brown podsol soil
that had been inoculated with eighteen *Eisenia foetida*. The dry matter yield of
the oat seedlings was 8.7% greater in the soil with earthworms than in that with
no earthworms.

In a series of greenhouse experiments, Atlavinyte and her co-workers (Atlavinyte et al. 1968, Atlavinyte 1974, Atlavinyte and Vanagas 1982) reported strong positive correlations between the numbers of earthworms added to soil (usually *A. caliginosa*) and the growth of barley. For instance, addition of 400 to 500 individuals of *A. caliginosa* m^{-2}, in one meter square field plots, increased the yield of barley by 78 to 96%. The increases in barley yields were proportional to the numbers of earthworms that had been added.

In field experiments in the Netherlands, large numbers of earthworms added to soil doubled the dry-matter yield of spring wheat, and increased grass yields fourfold and clover yields tenfold (van Rhee 1965). Kahsnitz (1922) reported that the addition of large numbers of live worms to a garden soil increased the yields of peas and oats by as much as 70%. Hopp and Slater (1948, 1949), working in the midwestern United States, reported that crop plants grown in a poorly-structured soil yielded 3160 kg ha^{-1} when earthworms were added to the soil at a rate of 120 m^{-2}, but only 280 kg ha^{-1} when no worms were added. These workers also investigated the influence of four different species of earthworms on crop yield, and reported that all four species caused consistent increases in yields of millet, lima beans, soybeans, and hay. They reported that the growth of soybeans and clover in soils with poor structure was stimulated more than that of grass and wheat. Dreidax (1931) reported that yields of winter wheat were greater in plots to which live worms were added than in plots with no worms. Uhlen (1953) showed that inoculation of soil with *L. terrestris* and *L. rubellus* increased yields of barley in heavily-manured garden soils.

In field experiments in New Zealand, the addition of introduced European species of lumbricid earthworms to pastures have increased grass yields consistently. Earthworms inoculated on a 10 m grid in a New Zealand pasture, spread evenly throughout the entire field in 7 to 8 years and increased yields of pasture grasses (Hamblyn and Dingwall 1945, Richards 1955, Stockdill 1959). There are many other instances of increased yields caused by earthworms in different parts of New Zealand (Nielson 1953, Waters 1955, Barley 1961, Barley and Kleinig 1964, Kleinig 1966, Noble et al. 1970), yields almost doubling in some instances (Stockdill 1982). Inoculation of worms to pastures with low earthworm populations is now a standard agricultural practice in New Zealand (see Lee, this volume).

Edwards and Lofty (1978) inoculated deep-burrowing species of lumbricids (*L. terrestris* and *A. longa*) into direct-drilled (no-till) field soils with low indigenous populations of earthworms. They compared cereal growth in plots to which the shallow-working species, *A. caliginosa* and *A. chlorotica*, had been added, with that in plots to which no earthworms had been added. They reported significantly improved root and shoot growth in response to the earthworm inoculations. In other field experiments (Edwards and Lofty 1980) added average populations of mixtures of either the deep burrowing species, *L. terrestris* and *A. longa*, or of the shallow-working species *A. caliginosa* and *A. chlorotica,* to small, replicated field plots. The growth of barley in inoculated plots was compared with that in plots to which no earthworms had been added. Both the numbers of barley plants and the rates of growth of the barley increased significantly in all of the inoculated plots, particularly in response to inoculation with deep-burrowing earthworm species.

Lavelle (1992) inoculated low numbers of selected earthworm species into low chemical input cropping systems at La Mancha, Mexico, Lamto in the Ivory Coast, and Yurimaguas, Peru. Crop growth and yields increased significantly in inoculated soils, in 10 out of 20 cropping cycles at all three sites. At La Mancha, when *Pontoscolex corethrurus* was introduced, 10 to 30% increases in the growth of grain crops occurred. At Yurimaguas, 145% increases in grain yields occurred. Different earthworm species may have different effects on plant growth. For example, Spain et al. (1992) reported that the growth of *Panicum maximum* increased significantly after inoculations with two species of eudrilid earthworms, *Chuniodrilus zielae* and *Stuhlmannia porifera*, but was unaffected by inoculation with the acanthodrilid worm, *Millsonia anomala*.

The knowledge that earthworms can improve soil fertility, has resulted in many attempts to add earthworms to poor soils in order to improve them or to use organic matter to encourage the build up of earthworm populations. Inoculation of earthworms into soil has been particularly promising in accelerating the reclaimation of flooded areas, such as Dutch polders, that were drained and put into cultivation (van Rhee 1962). For instance, the earthworms, *A. caliginosa and L. terrestris*, were introduced to polder soil at a rate of about 800 worms per tree, in soil planted with fruit trees. The trees grew more rapidly and exhibited greater root growth in the worm-inoculated soils than in those without worms (van Rhee 1969, 1971). The worms that were added multiplied rapidly in the polder soils; *A. caliginosa* increased from 4664 to 384,740 individuals ha^{-1} in 3 to 4 years, and *Allolobophora chlorotica* from 2588 to 12,666 individuals ha^{-1} over the same period (van Rhee 1969).

VII. Influences of Agricultural Practice on Earthworms

Earthworms have important influences on soil fertility and structure, so we need to know how they are affected by different agricultural practices. The four main management inputs into any farming system are cultivations, cropping patterns, fertilization, and crop protection. Each of these four inputs interacts strongly with the others. Thus, it is often difficult to extrapolate from the results of component research, which investigates the influence of one or another of these inputs on earthworms, to effects of whole systems on earthworm populations. In arid and semi-arid regions, irrigation is an additional major input that can have an important influence on earthworm populations.

A.   Cultivations

Although a single heavy cultivation does not usually have a drastic effect on earthworm populations, repeated cultivations over several seasons can depress earthworm populations progressively. Single cultivations can physically damage a proportion of the earthworm population, and can also destroy the upper parts of permanent burrows and expose any earthworms turned up on the soil surface to predation by birds and other predators (Edwards and Lofty 1978). There are also indirect effects of heavy cultivation on earthworms due to incorporation of surface litter into soil, loss of insulating cover, and redistribution of the organic food sources of the earthworms through the soil profile.

The kinds of cultivations used for crop production have changed dramatically over the past 30 to 40 years, from deep, moldboard plowing to various forms of conservation tillage, including chisel plowing and ridge-tillage, and to no tillage. No-till and most of the other conservation tillage practices, such as chisel-plowing and ridge tillage, all favor the buildup of larger earthworm populations, to levels limited only by climate and the availability of food (Gerard and Hay 1979, Barnes and Ellis 1982, Edwards and Lofty 1982a, House and Parmelee 1985). The deep-burrowing species, *L. terrestris*, *A. longa*, and *A. nocturna* benefit most from minimum cultivation. Chisel-plowing, shallow-tining, harrowing, and disking seem to have relatively small effects on either deep-burrowing or shallow-working species. The increases in earthworm populations that occur under long-term conservation tillage can be large. For example, Edwards and Lofty (1982b) reported a 30-fold difference between earthworm populations in plowed and no-till soils after 8 years.

B.   Fertilizers

Nearly all cropped soils in temperate regions are treated with either organic or inorganic fertilizers, all of which have an influence on earthworm populations. Marshall (1977) reviewed the overall effects of organic fertilizers on earthworm populations and reported that most of them increased earthworm abundance. For instance, Anderson (1983) and Curry (1976) reported that farmyard manures and animal slurries increased earthworm populations quite rapidly. The addition of a broad range of organic manures from sources such as cattle, hogs, poultry, municipal sewage wastes, and wastes from industries such as breweries, paper pulp, or potato processing can have a considerable influence on the buildup of earthworm populations in agricultural land (Edwards and Lofty 1979). Additions of some organic materials can double or triple earthworm populations in a single year. However, some liquid organic manures that have not been aged or composted can have short-term adverse effects on earthworm populations, due to their ammonia and salt contents; but populations usually recover quickly and thereafter increase.

Many inorganic fertilizers can also contribute indirectly to increases in earthworm populations, probably due to the increased amounts of crop residues returned to the soil (Zajonc 1975, Edwards and Lofty 1979, Barnes and Ellis 1982, Lofs-Holmin 1983a). Hendrix et al. (1992) reported that earthworm numbers in inorganically-fertilized meadows were, on average, nearly twice as numerous as those in unfertilized meadows on the Georgia piedmont in the southeastern United States. This was probably due to the increased plant growth, resulting in increased organic matter inputs, on the fertilized pastures. However, some inorganic fertilizers can adversely affect earthworms, which are very sensitive to ammonia, and ammonia-based fertilizers. Regular annual use of ammonium sulfate and anhydrous ammonia and sulfur-coated urea has been shown to decrease earthworm populations (Edwards and Lofty 1982b, Wei-Chun et al. 1990).

There is considerable evidence that liming or addition of fertilizer calcium helps to increase earthworm populations (Marshall 1977). For instance, Stockdill and Cossens (1966) showed that the addition of one ton of lime per acre to New Zealand soils caused a 50% increase in numbers of *A. caliginosa*. They reported that this process can be accentuated even further by inoculating the pastures with earthworms in spring when soils are moist, before adding lime. Lime can have direct effects on earthworms, by creating a more favorable microenvironment, influencing the pH, and providing Ca, which the earthworms need to survive and grow. Lime can also have indirect effects on earthworms by increasing crop growth, which results in a greater amount of organic residues that earthworms can use as food.

C.   Pesticides

The effects of pesticides on earthworms have been reviewed by Davey (1963), Edwards and Thompson (1973), and Edwards and Bohlen (1992). "Pesticides" is a generic term used for a group of compounds that includes: herbicides, fungicides, insecticides, acaricides, and nematicides. It is often assumed that most pesticides are toxic or otherwise harmful to earthworms. However, many pesticides are harmless or only slightly toxic to earthworms at normal application rates, and there is insufficient data on many pesticides, from field and laboratory assays, to make an accurate assessment of their relative toxicity. Most herbicides are not toxic and have little direct toxicity to earthworms. However, the majority of the triazine herbicides, such as atrazine, simazine, and cyanizine are slightly toxic and may have progressive adverse effects on earthworm populations, especially when they are used annually over extended periods of time (Edwards and Thompson 1973). More importantly, most herbicides can have drastic indirect effects on earthworms by increasing the availability of organic matter in the form of decomposing weeds (Edwards and Thompson 1973). Most fungicides have little direct effects on earthworms, with the exception of the carbamate-based fungicides such as carbendazim, benomyl, and thiophanate-methyl, which are all very toxic to earthworms. Of the insecticides in current use, the organophosphates, phorate, isozophos, chlorpyrifos, and ethoprophos, and most of the carbamate-based compounds such as carbaryl, carbofuran, methomy, and methiocarb are toxic to earthworms (Edwards and Bohlen 1992). All of the nematicides in current use that have been tested, whether fumigant or contact, such as D-D®, metham-sodium, and methyl bromide, have been reported to be toxic to earthworms.

The feeding and burrowing habits of different earthworm species can influence their susceptibility to pesticides. For instance, *L. terrestris* tends to be much more susceptible to pesticide residues on the soil surface or on surface crop residues, because of its habit of coming to the soil surface and moving over it in search of food. In certain situations, such as orchards, earthworms may be exposed to a combination of commonly used pesticides that are toxic to them, such as carbaryl, benomyl, and simazine; populations can be decreased significantly or completely eliminated in these situations. Some old orchards still have low earthworm populations from the historical use of inorganic and organochlorine pesticides (Edwards and Thompson 1973). There are few data in the literature on the chronic toxicity of pesticides to earthworms, other than on their impact on growth and cocoon production (Edwards and Bohlen 1992, Greg-Smith et al. 1992).

D.   Cropping Patterns

Cropping patterns can have considerable effects on earthworm populations in both space and time. However, there has been relatively little research into the effects of crop rotations on earthworm populations. One of the few detailed investigations of the influence of crop rotations on earthworm populations was by Lofs-Holmin (1983b). She concluded that the return of organic matter to the soil was the most important factor favoring the build-up of earthworm populations. Straw residues plowed into the soil and short-term hay crops greatly increased earthworm populations. In general, the inclusion of crops such as cereals, that leave considerable organic residues, encourages the buildup of earthworm populations much more than do legumes (e.g., soybean) which decompose quite rapidly and leave little residue (Edwards 1983). Perennial crops such as alfalfa, when included in a rotation, are particularly beneficial to the buildup of earthworm populations, partly because of the absence of tillage and partly because of the high protein content of the residues of these crops. Root crops, where most of the crop is removed, or periods of fallow both discourage the buildup of earthworm populations.

E.   Irrigation

Irrigated soils can support high levels of earthworm activity at times of year in any soil when moisture levels would normally be prohibitively dry to support earthworm activity (Barley and Kleinig 1964, Tisdal 1985, Blackwell and Blackwell 1989, Bezborodov and Khalbayeva 1990). Irrigation waters that carry earthworms or their cocoons may also be a source of inoculum for some earthworm species. The influence of earthworms on the capacity for water infiltration and minimization of runoff may be an important consideration in the successful management of irrigation systems (Kemper et al. 1987, Bezborodov and Khalbayeva 1990).

VIII.   Integrated Management of Earthworm Populations in Agroecosystems

Since the majority of evidence available indicates that earthworms improve crop productivity, it is important to design integrated management systems that promote the build up of earthworm populations. The key factor to such management seems to be the provision of adequate levels of fresh or decaying organic matter, since earthworms derive their nutrition from organic matter and associated microorganisms (Edwards and Fletcher 1988). Next in importance is to minimize mechanical disturbance by different forms of cultivation, which may harm earthworms directly or expose them to predators such as birds. Hence, the adoption of no-till or one of a range of conservation tillage practices is essential in maintaining abundant earthworm populations in agroecosystems. The use of all pesticides and fertilizers that have been reported as toxic to earthworms should be avoided, or they should be used only sparingly, and at infrequent intervals. Liming should be used in soils that are excessively acid, since a more neutral pH tends to favor the build up of earthworm populations of most species of earthworms. Changes in management practices alone may not be sufficient to build up earthworm populations even over a relatively long time period (50 years), particularly in soils with only low populations of one or a few species (Bohlen et al. 1994). In such situations, the introduction of other species may be necessary to ensure the build up of populations of those species that can provide maximum benefits to crop growth (see Lee, this volume).

IX.   Research Imperatives

Only in recent years has there been a significant increase in earthworm research in agroecosystems in North America. In the U.S. this may be due to the disciplinary nature of research, with earthworms not falling into any of the traditional disciplines. Furthermore, there has been a strong bias toward research on the physical and chemical aspects of soil science in the U.S., often to the exclusion of soil biology. Advances in our understanding of earthworms in agroecosystems will depend on new, integrated approaches to research. We have identified a number of priority areas for future research.

A.   Distribution and Abundance

Much more research is needed on the distribution of earthworm species in natural and managed ecosystems of North America. Currently, we lack a good understanding of the distribution and abundance of most species of earthworm in agroecosystems, and other managed ecosystems and in natural ecosystems in many regions. Some research emphasis needs to be placed on assessing the distribution of endemic species, which may be dominant in some agroecosystems and natural ecosystems but have been studied very little.

B.   Organic Matter Decomposition

There is a critical need for research into the effects of different earthworm species and communities on the breakdown, incorporation, and redistribution of organic matter in agricultural soils. This need is particularly acute in reduced and no-tillage agroecosystems and in those using mainly organic nutrient inputs. Moreover, there have been relatively few attempts to relate the abundance of different earthworm species to the rates of disappearance of different crop residues from the soil surface and incorporation and redistribution of these residues throughout the soil profile. Many models of crop residue decomposition do not take the important contribution of earthworms into account, despite the obvious role these organisms play. Earthworm effects need to be included explicitly in decomposition and soil organic matter models, especially in reduced and no tillage agroecosystems. Studies using crop residues labelled with stable or radioactive isotopes (e.g., ^{15}N, ^{13}C, ^{32}P) are needed to determine effects of earthworms on the transformation of nutrients derived from crop residues.

C.   Inoculation Studies

Investigations are needed into the effects of inoculating different earthworm species or species mixtures, into different soil types, on crop productivity. Laboratory experiments will continue to provide useful information on some aspects of this subject, but the most useful data must come from field experiments investigating the effects of large-scale manipulations of earthworm populations on crop growth and yield. Some of this research needs to focus specifically on the interactions between earthworm activities and root growth and turnover. Such research involves either adding earthworms to soils in which they are absent or, alternatively, comparing yields in plots with natural populations, with those in plots from which worms have been removed, either with electrical stimuli, biocides, or by formalin extraction. We are currently using both approaches to assess the importance of earthworms in field studies in Ohio (e.g., Blair et al. 1994). Such research needs to be long-term because many of the influences of earthworms on soils, such as those on macropore and aggregate formation, may take several years to develop after earthworms have been inoculated into soils, and may persist for several years after earthworms have been removed.

D.   Cast Production

Research quantifying the production of earthworm casts by earthworms with different ecological characteristics, and determining the distribution of casts in the soil profile, would help to assess the overall importance of earthworms on soil structure and nutrient availability.

E.   Cropping and Management Systems

The short- and long-term consequences of different cropping patterns and management systems on earthworm populations are poorly understood. Much more work in this area is needed before conscious attempts to manage earthworm populations can be integrated into overall farm management systems.

References

Aldag, R. and O. Graff. 1974. Einflub der Regenwurmtatigkeit auf proteingehalt und proteinqualitat junger Haferpflanzen. *Z. Landw. Forsch.* 31:277–284.

Allmaras, R.R., S.M. Copeland, J.F. Power, and D.L. Tanaka. 1994. Conservation tillage systems in the northernmost central United States. p. 256–287. In: Carter, M.R. (ed.) *Conservation tillage in temperate agroecosystems: development and adaptation to soil, climate, and biological constraints.* CRC Press, Boca Raton, FL.

Andersen, N.C. 1983. Nitrogen turnover by earthworms in arable plots treated with farmyard manure and slurry. p. 139–150. In: Satchell, J. E. (ed.) *Earthworm ecology: from Darwin to vermiculture.* Chapman and Hall, London.

Atlavinyte, O. 1974. Effect of earthworms on the biological productivity of barley. *Liet TSR Mokslu. Akad. Darb., Ser. C, Biol. Mokslai* 1:69–79.

Atlavinyte, O., Z. Bagdonaviciene, and L. Budviciene. 1968. The effect of Lumbricidae on the barley crops in various soils. *Pedobiologia* 8:415–423.

Atlavinyte, O. and J. Vanagas. 1982. The effect of earthworms on the quality of barley and rye grain. *Pedobiologia* 23:256–262.

Baluev, V.K. 1950. Earthworms of the basic soil types of the Iranov region. *Pochvovedenie* 487–491.

Barley, K.P. 1959a. The influence of earthworms on soil fertility. II. Consumption of soil and organic matter by the earthworm *Allolobophora caliginosa*. *Aust. J. Agric. Res.* 10:179–185.

Barley, K.P. 1959b. Earthworms and soil fertility. IV. The influence of earthworms on the physical properties of a red-brown earth. *Aust. J. Agric. Res.* 10:371–376.

Barley, K.P. 1961. The abundance of earthworms in agricultural land and their possible significance in agriculture. *Adv. Agron.* 13:249–268.

Barley, K.P. and C.R. Kleinig. 1964. The occupation of newly irrigated lands by earthworms. *Aust. J. Sci.* 26:290–291.

Barnes, B.T. and F.B. Ellis. 1982. The effects of different methods of cultivation and direct drilling, and of contrasting methods of straw disposal on populations of earthworms. *J. Soil Sci.* 30:669–679.

Basker, A., A.N. Macgregor, and J.H. Kirkman. 1992. Influence of soil ingestion by earthworms on the availability of potassium in soil: an incubation experiment. *Biol. Fertil. Soils* 14:300–303.

Bezborodov, G.A. and R.A. Khalbayeva. 1990. Effects of earthworms on the agrochemical and hydrophysical properties of irrigated sierozems. *Sov. Soil Sci.* 22:30–35.

Binet, F. and P. Trehen. 1992. Experimental microcosm study of the role of *Lumbricus terrestris* (Oligochaeta: Lumbricidae) on nitrogen dynamics in cultivated soils. *Soil Biol. Biochem.* 24:1501–1506

Blair, J.M., P.J. Bohlen, C.A. Edwards, B.R. Stinner, D.A. McCartney, and M.F. Allen. 1994. Manipulation of earthworm populations in field experiments in agroecosystems. *Acta Zool. Fennica* (in press).

Blanchart, E., P. Lavelle, and A.V. Spain. 1989. Effects of two species of tropical earthworms (Oligochaeta: Eudrilidae) on the size of aggregates in an African soil. *Rev. Écol. Biol. Sol* 26:417–425.

Blackwell, P.S. and J. Blackwell. 1989. The introduction of earthworms to an ameliorated, irrigated duplex soil in South-eastern Australia and the influence on macropores. *Aust. J. Soil. Res.* 27:807–814.

Bohlen, P.J., W.M. Edwards, and C.A. Edwards. 1994. Factors influencing earthworm community structure and diversity in experimental agricultural watersheds. *Plant Soil* (in press).

Bolton, P.J. and J. Phillipson. 1976. Burrowing, feeding, egestion and energy budgets of *Allolobophora rosea* (Savigny) (Lumbricidae). *Oecologia* 23:225–245.

Chadwick, L.C. and J. Bradley. 1948. An experimental study of the effects of earthworms on crop production. *Proc. Am. Soc. Hort. Sci.* 51:552–562.

Christensen, O. 1988. The direct effect of earthworms on nitrogen turnover in cultivated soils. *Ecol. Bull.(Stockholm)* 39:41–44.

Curry, J.P. 1976. Some effects of animal manures on earthworms in grassland. *Pedobiologia* 16:425–438.

Darwin, C. 1881. *The formation of vegetable mould through the action of worms: with observations on their habits*. Murray, London.

Davey, S.P. 1963. *Effects of chemicals on earthworms: a review of the literature*. Special Scientific Report on Wildlife, No. 74. U.S. Fish and Wildlife Service.

Doube, B.M., P.M. Stephens, C.W. Davoren, and M.H. Ryder. 1994. Interactions between earthworms, beneficial soil microorganisms and root pathogens. *J. Appl. Soil Ecol.* (in press).

C.A. Edwards, P.J. Bohlen, D.R. Linden, and S. Subler

Dreidax, L. 1931. Investigations on the importance of earthworms for plant growth. *Arch. Pflanzenbau.* 7:413–467.

Edwards, C.A. 1983. Earthworm ecology in cultivated soils. p. 123–137. In: Satchell, J.E. (ed.) *Earthworm ecology: from Darwin to vermiculture*. Chapman and Hall, London.

Edwards, C.A. and P.J. Bohlen. 1992. The effects of toxic chemicals on earthworms. *Rev. Environ. Contam. Toxic.* 125:24–99.

Edwards, C.A. and P.J. Bohlen. 1995. *The biology and ecology of earthworms*. 3rd ed. Chapman Hall, London (in press).

Edwards, C.A. and K.E. Fletcher. 1988. Interactions between earthworms and microorganisms in organic-matter breakdown. p. 235–248. In: Edwards, C.A., B.R. Stinner, D. Stinner, and S. Rabatin (eds.) *Biological interactions in soil*. Elsevier, Amsterdam, The Netherlands.

Edwards, C.A. and R. Lofty. 1977. *The biology of earthworms*. 2nd ed. Chapman Hall, London.

Edwards, C.A. and R. Lofty. 1978. The influence of arthropods and earthworms upon root growth of direct drilled cereals. *J. Appl. Ecol.* 15:789–795.

Edwards, C.A. and R. Lofty. 1979. The effects of straw residues and their disposal on the soil fauna. p. 37–44. In: Grossbard, E. (ed.) *Straw decay and its effect on dispersal and utilization*. John Wiley & Sons, New York.

Edwards, C.A. and R. Lofty. 1980. Effects of earthworm inoculation upon the root growth of direct drilled cereals. *J. Appl. Ecol.* 17:533–543.

Edwards, C.A. and R. Lofty. 1982a. Nitrogenous fertilizers and earthworm populations in agricultural soils. *Soil Biol. Biochem.* 14:515–521.

Edwards, C.A. and R. Lofty. 1982b. The effect of direct drilling and minimal cultivation on earthworm populations. *J. Appl. Ecol.* 19:723–734.

Edwards, C.A. and A.R. Thompson. 1973. Pesticides and the soil fauna. *Residue Rev.* 45:1–9.

Edwards, W.M., M.J. Shipitalo, W.A. Dick, and L.B. Owens. 1992. Rainfall intensity affects transport of water and chemicals through macropores in no-till soils. *Soil Sci. Soc. Am. J.* 56:52–58.

Edwards, W.M., M.J. Shipitalo, L.B. Owens, and L.D. Norton. 1989. Water and nitrate movement in earthworm burrows within long-term no-till corn fields. *J. Soil Water Cons.* 44:240–243.

Ehlers, W. 1975. Observations on earthworm channels and infiltration on a tilled and untilled loess soil. *Soil Sci.* 113:242–249.

Elliot, P.W., D. Knight, and J.M. Anderson. 1990. Denitrification in earthworm casts and soil from pasture under different fertilizer and drainage regimes. *Soil Biol. Biochem.* 22:601–605.

Gerard, B.M. and R.K.M. Hay. 1979. The effects on earthworms of ploughing, tined cultivation, direct drilling and nitrogen in a barley monoculture system. *J. Agric. Sci. (Cambridge)* 93:147–155.

Graff, O. 1971. Stickstoff, Phosphor und Kalium in der Regenwurmlosung auf der Wiesenversuchsfläche des Sollinprojektes. p. 503–511. In: D'Aguilar, J. (ed.) *IV Colloquium Pedobiologiae*. Institute National des Recherches Agriculturelles Publication 71-7, Paris.

Greig-Smith, P.W., H. Becker, P.J. Edwards, and F. Heimbach. 1992. *Ecotoxicology of earthworms*. Intercept Publishing, Andover, England. 269 pp.

Hamblyn, C.J. and A.R. Dingwall. 1945. Earthworms. *N.Z. J. Agric.* 71:55–58.

Hendrix, P.F., B.R. Mueller, R.R. Bruce, G.W. Langdale, and R.W. Parmelee. 1992. Abundance and distribution of earthworms in relation to landscape factors on the Georgia piedmont, U.S.A. *Soil Biol. Biochem.* 24:1357–1361.

Hoeksema, K.J. and A. Jongerius. 1959. On the influence of earthworms on the soil structure of mulched orchards. *Proc. Int. Symp. Soil Struct. Ghent* 1958, pp. 188–194.

Holmes, G.A. 1952. Molybdenum responses at Invermay. *Proc. N.Z. Grassland Assoc.* pp. 198–201.

Hopp, H. 1973. *What every gardener should know about earthworms.* Garden Way Publishing Company, VT.

Hopp, H. and H.T. Hopkins. 1946. Earthworms as a factor in the formation of water-stable aggregates. *J. Soil Water Cons.* 1:11–13.

Hopp, H. and C.S. Slater. 1948. Influence of earthworms on soil productivity. *Soil Sci.* 66:421–428.

Hopp, H. and C.S. Slater. 1949. The effect of earthworms on the productivity of agricultural soils. *J. Agric. Res.* 78:325–339.

House, G.J. and R.W. Parmelee. 1985. Comparison of soil arthropods and earthworms from conventional and no-tillage agroecosystems. *Soil and Tillage Res.* 5:351–360.

James, S.W. 1991 Soil, nitrogen, phosphorus, and organic matter processing by earthworms in tallgrass prairie. *Ecology* 76:2101–2109.

Jeanson, C. 1964. Micromorphology and experimental soil zoology: contribution to the study, by means of giant-sized thin sections, of earthworm-produced artificial soil structure. p. 47–55. In: Jongerius, A. (ed.) *Soil micromorphology.* Proceedings of the 2nd International Work Meeting of Soil Micromorphology, Arnhem, The Netherlands.

Joshi, N.V. and B.V. Kelkar. 1952. The role of earthworms in soil fertility. *Indian J. Agric. Sci.* 22:189–196.

Kahsnitz, H.G. 1922. Investigations on the influence of earthworms on soil and plant. *Botanical Archives* 1:315–351.

Kemper, W.D., T.J. Trout, A. Segeren, and M. Bullock. 1987. Worms and water. *J. Soil Water Cons.* 42:401–404.

Kladivko, E.J., A.D. Mackay, and J.M. Bradford. 1986. Earthworms as a factor in the reduction of soil crusting. *Soil Sci. Soc. Am. J.* 50:191–196.

Kleinig, C.R. 1966. Mats of unincorporated organic matter under irrigated pasture. *Aust. J. Agric. Res.* 17:327–333.

Knight, D., P.W. Elliot, J.M. Anderson, and D. Scholefield. 1992. The role of earthworms in managed, permanent pastures in Devon, England. *Soil Biol. Biochem.* 24:1511–1517.

Kretzchmar, A. 1978. Quantification écologique des galeries de lombriciens. Techniques et première estimations. *Pedobiologia* 18:31–38.

Kretzchmar, A. 1982. Description des galeries des vers de terre et variation saisonnières des réseaux (observations en conditions naturelles). *Rev. Écol. Biol. Sol* 19:579–591.

Lavelle, P. 1988. Earthworm activities and the soil system. *Biol. Fertil. Soils* 6:237–251.

Lavelle, P. 1992. *Conservation of soil fertility in low-input agricultural systems of the humic tropics by manipulating earthworm communities (macrofauna project).* European Economic Community Project No. TS2-0292-F (EDB), pp. 138.

Lavelle, P. and C. Fragoso. 1992. Earthworm communities of tropical rainforests. *Soil Biol. Biochem.* 24:1397–1408.

Lavelle, P. and A. Martin. 1992. Small-scale and large-scale effects of endogeic earthworms on soil organic matter dynamics in soil of the humid tropics. *Soil Biol. Biochem.* 24:1491–1498.

Lee, K.E. 1983. The influence of earthworms and termites on nitrogen cycling. p. 35–48. In: Lebrun, Ph., H.M. André, and G. Wauthy. (eds.) *New trends in soil biology.* Proceedings of the 8th International Colloquium on Soil Zoology, Dieu-Brichart, Ottignies-Louvain-la-Neuve.

Lee, K.E. 1985. *Earthworms: their ecology and relationships with soils and land use.* Academic Press, New York.

Lee, K.E. and R.C. Foster. 1992. Soil fauna and soil structure. *Aust. J. Soil. Res.* 29:745–775.

Lofs-Holmin, A. 1983a. Influence of agricultural practices on earthworms (Lumbricidae). *Acta Agriculturae Scandinavica* 33:225–234.

Lofs-Holmin, A. 1983b. Earthworm population dynamics in different agricultural rotations. p. 151–160. In: Satchell, J.E. (ed.) *Earthworm ecology: from Darwin to vermiculture.* Chapman and Hall, London.

Logsdon, D.R. and D.R. Linden. 1992. Microrelief and rainfall effects on water and solute movement in earthworm burrows. *Soil Sci. Soc. Am. J.* 56:727–733.

Lunt, H.A. and G.M. Jacobson 1944. The chemical composition of earthworms casts. *Soil Sci.* 58:367–375.

Mackay, A.D. and E.J. Kladivko. 1985. Earthworms and rate of breakdown of soybean and maize residues in soil. *Soil Biol. Biochem.* 17:851–857.

Mackay, A.D., J.K. Syers, J.A. Springett, and P.E.H. Gregg. 1982. Plant availability of phosphorous in superphosphate and a superphosphate rock as influenced by earthworms. *Soil Biol. Biochem.* 14:281–287.

Madsen, E.L. and M. Alexander. 1982. Transport of *Rhizobium* and *Pseudomonas* through soil. *Soil Sci. Soc. Am. J.* 46:557–560.

Mansell, G.P., J.K. Syers, and P.E.H. Gregg. 1981. Plant availability of phosphorous in dead herbage ingested by surface-casting earthworms. *Soil Biol. Biochem.* 13:163–167.

Marinissen, J.Y.C. and A.R. Dexter. 1990. Mechanisms of stabilization of earthworm casts and artificial casts. *Biol. Fertil. Soils* 9:163–167.

Marshall, V.G. 1977. *Effects of manures and fertilizers on soil fauna: a review.* Commonwealth Bureau of Soils, 79 pp.

Martin, A. 1991. Short- and long-term effects of the endogeic earthworm *Millsonia anomala* (Omodeo) (Megascolicidae, Oligochaeta) of tropical savannas, on soil organic matter. *Biol. Fertil. Soils* 11:234–238.

Mather, J.C. and O. Christensen. 1988. Surface movements of earthworms in agricultural land. *Pedobiologia* 32:399–405.

McColl, H.P., P.B.S. Hart, and F.J. Cook. 1982. Influence of earthworms on some soil chemical and physical properties, and the growth of ryegrass on a soil after topsoil stripping—a pot experiment. *N.Z. J. Agric. Res.* 25:239–243.

Mulongoy, K. and A. Bedoret. 1989. Properties of worm casts and surface soils under various plant covers in the humid tropics. *Soil Biol. Biochem.* 21:197–203.

Nielson, R.L. 1953. Earthworms (recent research work). *N.Z. J. Agric.* 86:374.

Noble, J.C., W.T. Gordon, and C.R. Kleinig. 1970. The influence of earthworms on the development of mats of organic matter under irrigated pasture in Southern Australia. *Proceedings of the 11th International Grassland Congress*, pp. 465–468.

Oades, J.M. 1993. The role of biology in the formation, stabilization and degradation of soil structure. *Geoderma* 56:377–400.

Parle, J.N. 1963. Microorganisms in the intestines of earthworms. *J. Gen. Microbiol.* 31:1–11.

Parmelee, R.W. and D.A. Crossley, Jr. 1988. Earthworm production and role in the nitrogen cycle of a no-tillage agroecosystem on the Georgia piedmont. *Pedobiologia* 32:351–361.

Parmelee, R.W., M.H. Beare, W. Cheng, P.F. Hendrix, S.J. Rider, D.A. Crossley, Jr., and D.C. Coleman. 1990. Earthworms and enchytraeids in conventional and no-tillage agroecosystems: a biocide approach to assess their role in organic matter breakdown. *Biol. Fertil. Soils* 10:1–10.

Piearce, T.G. 1978. Gut contents of some lumbricid earthworms. *Pedobiologia* 18:153–157.

Potter, D.A., M.C. Buxton, C.T. Redmond, C.G. Patterson, and A.J. Powell. 1990. Toxicity of pesticides to earthworms (Oligochaeta: Lumbricidae) and effect on thatch degradation in Kentucky bluegrass turf. *J. Econ. Entom.* 83: 2362–2369.

Raw, F. 1962. Studies of earthworm populations in orchards I. Leaf burial in apple orchards. *Annals Appl. Biol.* 50:389–404.

Reddell, P. and A.V. Spain. 1991. Earthworms as vectors of viable propagules of mycorrhizal fungi. *Soil Biol. Biochem.* 23:767–774.

Richards, J.G. 1955. Earthworms (recent research work) *N. Z. J. Agric.* 91:559.

Rouelle, J. 1983. Introduction of amoebae and *Rhizobium japonicum* into the gut of *Eisenia fetida* (Sav.) and *Lumbricus terrestris* L. p. 372–381. In: Satchell, J.E. (ed.) *Earthworm ecology: from Darwin to vermiculture.* Chapman and Hall, London.

Satchell, J.E. 1967. Lumbricidae. p. 259–322. In: Burges, A. and F. Raw (eds.) *Soil biology.* Academic Press, London.

Scheu, S. 1987. Microbial activity and nutrient dynamics in earthworm casts (Lumbricid-ae). *Biol. Fertil. Soils* 5:230–234.

Scheu, S. 1991. Mucus secretion and carbon turnover of endogeic earthworms. *Biol. Fertil. Soils* 12:217–220.

Sharpley, A.N. and J.K. Syers. 1976. Potential role of earthworm casts for the phospho-rous enrichment of runoff waters. *Soil Biol. Biochem.* 8:341–346.

Sharpley, A.N., J.K. Syers, and J.A. Springett. 1979. Effect of surface-casting earthworms on the transport of phosphorous and nitrogen in surface runoff from pasture. *Soil Biol. Biochem.* 11:459–462.

Shaw, C. and S. Pawluk. 1986a. Faecal microbiology of *Octolasion tyrtaeum*, *Aporrectodea turgida* and *Lumbricus terrestris* and its relation to carbon budgets of three artificial soils. *Pedobiologia* 29:377–389.

Shaw, C. and S. Pawluk. 1986b. The development of soil structure by *Octolasion tyrtaeum*, *Aporrectodea turgida* and *Lumbricus terrestris* in parent materials belonging to different textural classes. *Pedobiologia* 29:327–339.

Shipitalo, M.J., W.M. Edwards, and C.E. Redmond. 1994. Comparison of water movement and quality in earthworm burrows and pan lysimeters. *J. Environ. Qual.* 23:1345–1351.

Shipitalo, M.J. and R. Protz. 1988. Factors influencing the dispersibility of clay in worm casts. *Soil Sci. Soc. Am. J.* 52:764–769.

Spain, A.V., P. Lavelle, and A. Mariotti. 1992. Stimulation of plant growth by tropical earthworms. *Soil Biol. Biochem.* 24:1629–1633.

Springett, J.A. and J.K. Syers. 1979. The effect of earthworm casts on ryegrass seedlings. p. 44–47. In: Crosby, T.K. and R.P. Pottinger (eds.) *Proc. 2nd Australasian Conf. Grassland Invert. Ecol.* Government Printer, Wellington, New Zealand.

Stephens, P.M., C.W. Davoren, B.M. Doube, and H. Ryder. 1994. Ability of earthworms *Aporrectodea rosea* and *Aporrectodea trapezoides* to increase plant growth and the foliar concentration of elements in wheat (*Triticum aestivum* cv. Spear) in a sandy loam soil. *Biol. Fertil. Soils* (in press).

Stewart, V.I., L.O. Salih, K.H. Al-Bakri, and J. Strong. 1980. Earthworms and soil structure. *Welsh Soils Discussion Group Report* 20:103–114.

Stockdill, S.M.J. 1959. Earthworms improve pasture growth. *N.Z. J. Agric.* 98:227–233.

Stockdill, S.M.J. 1966. The effect of earthworms on pastures. *Proc. N.Z. Ecol. Soc.* 13:68–75.

Stockdill, S.M.J. 1982. Effects of introduced earthworms on the productivity of New Zealand pastures. *Pedobiologia* 24:29–35.

Stockdill, S.M.J. and G.G. Cossens 1966. The role of earthworms in pasture production and moisture conservation. *Proc. N.Z. Ecol. Soc.* 13:68–74.

Svensson, B.H., U. Boström, and L. Klemedston. 1986. Potential for higher rates of denitrification in earthworm casts than in the surrounding soil. *Biol. Fertil. Soils* 2:147–149.

Swaby, R.J. 1950. The influence of earthworms on soil aggregation. *J. Soil Sci.* 1:195–197.

Syers, J.K., A.N. Sharpley, and D.R. Keeney. 1979. Cycling of nitrogen by surface-casting earthworms in a pasture ecosystem. *Soil Biol. Biochem.* 11:181–185.

Syers, J.K. and J.A. Springett. 1984. Earthworms and soil fertility. *Plant Soil* 76:93–104.

Teotia, S.P., F.L. Duley, and T.M. McCalla. 1950. Effect of stubble mulching on number and activity of earthworms. *Neb. Agr. Exp. Stn. Res. Bull.* 165:1–20.

Tisdall, J.M. 1985. Earthworm activity in irrigated red-brown earths used for annual crops in Victoria. *Aust. J. Agric. Res.* 23:291–299.

Tisdall, J.M. and J.M. Oades. 1982. Organic matter and water-stable aggregates in soils. *J. Soil Sci.* 33:141–163.

Tiwari, S.C., B.K. Tiwari, and R.R. Mishra. 1989. Microbial populations, enzyme activities and nitrogen-phosphorous-potassium enrichment in earthworm casts and in the surrounding soil of a pineapple plantation. *Biol. Fertil. Soils* 8:178–182.

Tomati, U., A. Grappelli, and E. Galli. 1988. The hormone-like effect of earthworm casts on plant growth. *Biol. Fertil. Soils* 5:288–294.

Trojan, M.D. and D.R. Linden. 1992. Microrelief and rainfall effects on water and solute movement in earthworm burrows. *Soil Sci. Soc. Am. J.* 56:727–733.

Uhlen, G. 1953. Preliminary experiments with earthworms. *Landbr. Hogsk. Inst. Jordkultur Meld.* 37:161–183.

van Rhee, J.A. 1962. Inoculation of earthworms in a newly drained polder. *Pedobiologia* 9:128–132.

van Rhee, J.A. 1965. Earthworm activity and plant growth in artificial cultures. *Plant Soil* 22:45–48.

van Rhee, J.A. 1969. Development of earthworm populations in polder soils. *Pedobiologia* 9:128–140.

van Rhee, J.A. 1971. Some aspects of the productivity of orchards in relation to earthworm activities. *Ann. Zool. Ecol. Anim. Spec. Publ.* 4:99–108.

van Rhee, J.A. 1977. A study of the effect of earthworms on orchard productivity. *Pedobiologia* 17:107–114.

Waters, R.A.S. 1955. Numbers and weights of earthworms under a highly productive pasture. *N. Z. J. Sci. Tech.* A36:516–525.

Wei-Chun, Ma, L. Brussaard, and J.A. DeRidder. 1990. Long-term effects of nitrogenous fertilizers on grassland earthworms and their relation to soil acidification. *Agric. Ecosys. Environ.* 30:71–80.

Wollny, E. 1890. Untersuchungen über die Beeinflussung der Fruchtbarkeit der Ackerkrume durch die Tatigkeit der Regenwürmer. *Forschn Geb. AgrikPhys. Bodenk.* 13:381–395.

Wolters, V. and R.G. Joergensen. 1992. Microbial carbon turnover in beech forest soil worked by *Aporrectodea caliginosa* (Savigny) (Oligochaeta: Lumbricidae). *Soil Biol. Biochem.* 24:171–177.

Zachman, J.E. and D.R. Linden. 1989. Earthworm effects on corn residue breakdown and infiltration. *Soil Sci. Soc. Am. J.* 53:1846–1849.

Zachman, J.E., D.R. Linden, and C.E. Clapp. 1987. Macroporous infiltration and redistribution as affected by earthworms, tillage and residue. *Soil Sci. Soc. Am. J.* 51:1580–1586.

Zajonc, I. 1975. Variation in meadow associations of earthworms caused by the influence of nitrogen fertilizers and liquid manure irrigation. p. 497–503. In: Vanek, J. (ed.) *Progress in soil zoology.* Dr. W. V., The Hague, The Netherlands.

Earthworms and Sustainable Land Use
K.E. Lee

I. Introduction

It is not easy to define sustainable land use, nor to recognize it if it is attained. It is easier to recognize degradation of soils that indicate land use practices that are not sustainable and to modify management practices to counter further degradation and perhaps to repair damage done by past misuse. The beneficial effects on soils that are known to derive from earthworm activity, such as effects on soil structure, water infiltration, aeration, organic matter incorporation, nutrient availability, and nutrient cycling, are widely recognized, as discussed in previous chapters in this volume. Management of earthworm populations can be directed towards maximizing these beneficial effects and so to contribute to sustainability in land use. Management intervention may take the form of promoting the activities of existing earthworm species associations, sometimes resulting in changes in the proportions of their component species,

1-56670-053-1/95/$0.00+$.50
©1995 by CRC Press, Inc.

or it may involve the introduction of additional species to add to or enhance the effects of existing species associations.

Intelligent management of earthworm populations depends on an understanding of the diversity of earthworm communities, the nature and significance of species associations, the ecology and behavior of individual species, and the role of earthworms as contributors to the biological processes that are a fundamental component of soil formation and soil fertility. It also requires substantial knowledge of pedology, soil physics, and soil chemistry, especially as they affect and are affected by biological processes.

Soils are the products of a combination of *abiotic factors*, including the composition of the parent rock, physical and chemical rock weathering, temperature regimes, hydrology and landscape stability, and *biotic factors*, especially uptake by plants of water and nutrients, photosynthesis, and the decomposition of dead plant material and recycling of its constituents (Lee 1983). Feedback processes operate within and between the abiotic and the biotic processes, such that "downstream consumers" increase the rate of production of "upstream producers" and contribute critically to the over all nature and rates of large scale geochemical, hydrological, and climatic cycles (Odum and Biever 1984, van Breemen 1993). This concept has overtones of Lovelock's Gaia hypothesis, which can not be uncritically accepted. Its application to earthworms was clearly recognized by Darwin (1837, 1881). As in his theories of evolution and of the formation of coral reefs, he showed with earthworms how the accumulation of many small qualitative causes can produce great quantitative effects and, in the particular case of earthworms, how the sum of the activities of the very large numbers present in many soils has a major influence on soil formation and soil fertility.

Brun et al. (1987) reviewed the literature on the use of earthworms to stimulate soil fertility. They listed 73 papers that deal specifically with various aspects of the subject in Europe, the former USSR, North America, and Australasia, concluding that the beneficial effects of earthworms are well known, but that there is an urgent need for the development of methodology for broad scale application of this knowledge to the better management of soils.

Some soils have developed and carried natural woodland or prairie vegetation over long periods without earthworms. For example, when the first European settlers arrived in regions of North America that were covered by the ice sheets of Pleistocene and Recent geological times, and also in large portions of the Great Basin and Great Plains that were not glaciated, there were forests and prairies on soils that had no earthworms (Gates 1970). Earthworms that are now found in the farmlands of these regions are common European lumbricids that were apparently introduced accidentally as settlement and agricultural development proceeded (Gates 1970, 1976, Lee 1985, p. 159).

Nutrient and energy flows and food webs in the ecosystems that predated farming must have achieved an equilibrium, such that inputs to plants of nutrients weathered from soil minerals, and fixation of carbon and nitrogen from the atmosphere, were more or less balanced by outputs through leaching losses and decomposition processes. Use of the soil for food production disturbs this equilibrium. Food production involves cultivation, the elimination of previously dominant plant species, their replacement with new dominants, and a general reduction in plant diversity; harvesting of crops, forestry, or feeding of grazing animals removes from the ecosystem a proportion of the energy and plant nutrients captured in plant growth. Associated changes in the nature of plant litter, the distribution of plant roots, the physical and chemical properties of soils, and the geometry of aggregates and the spaces between them, impact upon the soil biota. This results generally in reduced diversity, although total numbers of soil organisms may not be reduced. Opportunities become available for invasion or introduction of species that can occupy niches that did not exist previously. It is evident from work in New Zealand, the Netherlands, and elsewhere (see review in Lee 1985) that introduction of earthworms to agricultural lands where they are absent, or modification and management of earthworm communities, including introduction of carefully selected species, can be particularly important in promoting plant production in agroecosystems.

The beneficial effects of earthworms on physical and chemical characteristics of soils have been outlined by other contributors to this volume. It is not suggested that earthworms add any nutrients, but their activities, digestive processes and associations with microorganisms result in increased availability and rates of cycling of plant nutrients, and changes in soil physical properties that promote plant growth.

This paper deals briefly with the diversity of earthworms, especially diversity within communities, and with concepts of management of earthworm populations to promote sustainable soil productivity.

II.   Diversity

A.   Species Associations

The number of species that make up a community is a simple measure of diversity, basic to the consideration of niche-partitioning and sharing of resources among species. Lee (1985) listed numbers of species of earthworms recorded from a wide variety of locations and vegetation types. Some examples are in Table 1. The range is from one to fifteen species, rarely more than nine, and most commonly two to five, with a remarkable consistency that is more or less independent of taxonomic groups and major geographical regions.

Lumbricid associations in Europe include fewer species in northern than in warmer southern latitudes, are relatively species-poor in peat lands, heath lands, coniferous forests, and arable lands, and relatively species-rich in deciduous forests, permanent pastures, and gardens. The species-poor northern species associations usually include *Dendrobaena octaedra* (Savigny) and/or *Dendro-drilus rubidus* (Savigny), which are small and confined to litter or superficial layers of the predominantly acidic podzols and other soils of northern latitudes.

Table 1. Number of species in earthworm communities from various habitats and geographical regions. Data from Lee (1985) and J.W. Reynolds (pers. comm.).

Location	Vegetation Type	Number of Species[a]	Taxonomic Groups
Canada/northern U.S.	Forest and grassland	3–4	Lumbricidae
Central U.S.	Deciduous forests	4–5	Lumbricidae
	Coniferous forests	3–4	Megascolecidae
	Grasslands	4	Komarekionidae
Southern U.S.	Coniferous forests	2–3	Lumbricidae and Megascolecidae
South America	Tropical rainforest	8	Glossoscolecidae
Scandinavia	Heath	2	
	Coniferous forest	3–4	Lumbricidae
	Deciduous forest	8–9	
	Pastures and meadows	5–7	
Western, central, and southern Europe	Heath	2–7	
	Deciduous forest	6–9	Lumbricidae
	Pastures and meadows	5–15	
Africa	Savanna	4–9	Eudrilidae and Megascolecidae
	Forest	4–11	
Australia	Pasture	2	Lumbricidae
	Woodland	2–6	Megascolecidae
	Alpine herbfield	2–5	
	Pastures	2–5	Lumbricidae and Megascolecidae
	Sugar cane	1	Glossoscolecidae
New Zealand	Rainforest	1–7	Megascolecidae (and Lumbricidae)
	Native shrubland	1–3	
	Pine plantations	1–3	
	Native grasslands	1–5	
	Pastures	1–4	Lumbricidae (and Megascolecidae)

[a]Number of species are means or ranges of various estimates.

Southern European species associations are more diverse, including increased proportions of larger, soil-dwelling species of *Allolobophora* s.l. and *Lumbricus* spp. in the less acidic brown forest soils, calcareous soils, etc., of more southern latitudes.

Lavelle (1983) contrasted the species-poorness of earthworm populations with the species-richness of insect and mite populations in soils; he commented on the lack of increasing diversity in a meridional sequence of populations from Sweden to the Ivory Coast and contrasted this with the general rule, confirmed for many animal groups, of increasing diversity in such meridional sequences (MacArthur and Wilson 1967). The species in the cold northern regions are small and there is an increasing diversity of size that runs through the meridional sequence, with the widest range in tropical regions. Lavelle (1983) proposed that the increasing diversity of size, and therefore of the functional diversity of earthworms in the ecosystem, corresponds to the more general increasing species-diversity of other animal groups. The idea that diversity of size increases meridionally does not hold for the megascolecid earthworms of Australasia and Oceania, where very large species and great diversity of size are common in the south, while associations of small and medium sized species, especially of the *Pheretima* s.l. group, are most widespread in tropical regions (Lee 1985).

B.   Functional Groups

The diversity of earthworms might best be described in terms of functional groups, rather than of individual species. Functional groups of earthworms are made up of individuals or species, with or without close taxonomic affinities, that share common morphological features, reflected in a variety of behavioral, trophic, and other characteristics. "Functional" attributes are likely to provide more information about organisms' responses to environmental change than is available in a species' name or taxonomic description. They can also be used as a basis for assessing the potential of individual species and species associations to affect the physical and chemical characteristics of soils, and so are useful in selecting candidate species for introduction programs.

Lee (1959) recognized three major functional groups among the megascolecid earthworms of New Zealand (Figure 1). He distinguished them primarily in relation to the soil horizons in which they are most commonly found, into leaf mould (litter or O-horizon) species, topsoil (A-horizon) species, and subsoil (B- or B/C-horizon) species, and further defined these groups on the basis of shared morphological, behavioral and physiological features, their food preferences, susceptibility to predation, geographical distribution, and reaction to changes in land use patterns.

Bouché (1972, 1977) independently of Lee (1959) recognized three major morpho-ecological groups among European lumbricids (Figure 2). He distinguished the groups primarily on the basis of morphological characters that he considered to have functional significance, and correlated these characters with behavioral features. The resultant groups were remarkably similar to Lee's

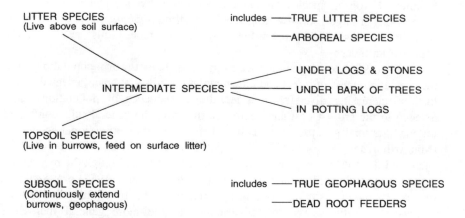

Figure 1. Functional groups of earthworms recognized by Lee (1959), based on New Zealand Megascolecoidea.

groups of megascolecids. As defined in his paper of 1977, Bouché's groups were: epigeic species, that live above the mineral soil surface, typically in the litter layers of forest soils; anecic species, that live in burrows in the mineral soil, but come to the surface to feed on dead leaves, which they drag into their burrows; endogeic species, that live in the mineral soil horizons, burrowing continuously and feeding on soil more or less enriched with organic matter.

The groups recognized by Lee and Bouché, based as they are on different parameters and applying them independently to taxonomically and geographically remote species associations, give some assurance that the groups have some ecological significance. Similarities extend beyond the characters that were initially used in defining the groups; some correlates are listed in Table 2. Their

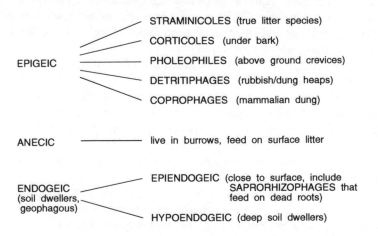

Figure 2. Functional groups of earthworms recognized by Bouché (1972, 1977), based on European Lumbricidae.

Table 2. Some characters that correlate with the major functional groups recognized in New Zealand Megascolecoidea by Lee (1959) and in European Lumbricidae by Bouché (1972, 1977)

MORPHOLOGICAL	Size
	Pigmentation
	Muscle development
	Gut structure
BEHAVIORAL	Form of burrows
	Casting behavior
	Mobility
	Reaction to desiccation
DEMOGRAPHIC	Cocoon production
	Generation time
	Length of life
ENVIRONMENTAL	Food supply
	Predatory pressure
	Effect of changes
GEOGRAPHICAL RANGE OF SPECIES	

significance has been further confirmed by the recognition of the same or similar functional groups of earthworms by researchers in many parts of the world, e.g., in Sweden (Nordström and Rundgren 1974), England (Phillipson et al. 1976), Venezuela (Németh and Herrera 1982), western Africa (Lavelle 1983), and North America (James and Cunningham 1989).

   Lavelle (1979) further developed the concept of functional groups that had been proposed by Lee and Bouché in his assessment of species associations from the savannah soils of Ivory Coast. He proposed that selection may have favored the development of distinct demographic profiles, characteristic of species best fitted to distinct subdivisions of the soil environment. He described these profiles in terms of three parameters: (a) total duration (i.e., maximum) of growth period; (b) life expectancy at hatching; and (c) number of cocoons produced per adult worm. Demographic data from Ivory Coast species showed that the duration of growth is shorter for small species that live close to the surface than for deeper dwelling species, that life expectancy at hatching is directly related to mean maximum weight and mean depth of activity, and that number of cocoons per year is at a maximum for certain small species and at a minimum for large geophagous species. He proposed a formula, to apply to individual species

$$D = 10^3 \cdot F/(C \cdot Ev)$$

where F = number of cocoons per year, C = duration of growth in months, Ev = life expectancy (in months) at hatching, and D = a demographic index that provides a measure of the capability of a species for population growth. Lavelle

(1979) further extended his concept of demographic profiles by examining the relationship between the demographic index and energy utilization of individual species. The value of D was shown to correlate closely with total tissue production, mean biomass, and the proportion of production devoted to reproduction, i.e., species whose production per unit biomass is high invest more energy in reproduction than those with low production per unit biomass. On this basis Lavelle further showed that species with high potential for population growth take advantage of food resources that are relatively high in nutritive value (litter) but are not permanently available, while species with low potential for population growth take advantage of food resources with low nutritive value (soil) but with reliable availability. Lavelle (1979) related the demographic strategies that he recognized in these two groups of earthworms to the more general existence of similar ecological strategies recognized by Pianka (1970) and Calow (1977).

Demographic strategies provide a link between the vertical stratification of the functional groups recognized by Lee (1959) and Bouché (1972, 1977) and the ecological energetics of individual species that make up species associations (see Lee 1985). This provides a means of comparing the functional significance of the species that make up one community with that of the species in another community, and so to assess the possible advantages for soil improvement and plant production to be gained from management of existing species associations or introduction of species. It is necessary to have some detailed knowledge of the phenology and behavior of a species to determine a demographic index and there are few species for which sufficient information is available. Most earthworm communities, however, include only a few species, and this is particularly so of communities in farm lands.

C.   Genetic Diversity

The niche width of a particular species, its behavior and its relationships with other species and with abiotic environmental factors, tend to be treated as constant, even when the species is in another community of different species composition in another ecosystem. This gives too narrow a view of the functional role of earthworm species. Their ecological plasticity (see below) points to the possible significance of intraspecific genetic diversity, especially among peregrine species.

Polyploidy is common in earthworms and is well known among the most widespread of peregrine lumbricids (Lee 1987). Omodeo (1952, 1953) found 13 polyploid races among 29 species of lumbricids. Parthenogenesis is obligatory for most polyploid races of earthworms, because spermatogenesis is unsuccessful. There is an obvious advantage for obligate parthenogenetic forms, compared with diploid, bisexual forms, in achieving widespread dispersal in new habitats, and Omodeo (1952) noted that, within species that include parthenogenetic races, polyploids were more widespread than were bisexual diploids. Jaenike et al.

(1982) found a number of clones of parthenogenetic pentaploid *Dendrodrilus rubidus* in northeastern United States. They noted that pentaploid races of this species are not known from Europe, indicating that the species might be native to North America as well as to Europe. Lee (1987) pointed out that there may have been pentaploids among the worms that were introduced and that they are now dominant because they have an advantage over diploids.

Preferential selection from initial introduced populations, based on genetic heterogeneity, may be inferred from the temperature preferenda of peregrine lumbricids. For example, the optimum temperature preference of field populations of *Aporrectodea rosea* (Savigny) in Germany was shown by Graff (1953) to be 12°C, while Reinecke (1975) showed that this species in South Africa had an optimum range of 25.0 to 26.9°C in laboratory cultures. There is an obvious advantage for South African populations in their tolerance for higher temperatures than those preferred by European populations. Many other behavioral and physiological differences that may indicate the effects of selective pressures on genetic variations in the original stock are apparent between widely separated populations of peregrine species.

DNA fingerprinting techniques are being used in southern Australia (G.H. Baker, in press) for recognition of strains of *Aporrectodea trapezoides* (Dugès) and *A. caliginosa* [probably the same as *A. turgida* (Eisen) in U.S. usage]. It seems likely that the original stock of these species came to Australia with plants that were brought, mainly from Great Britain, in barrels of soil. The DNA fingerprinting technique will be applied to worms from Australia and from Europe with a view to tracing their origins in Europe and to selecting strains that are particularly suited to Australian environments, either from Australia or from Europe, to be used in introduction programs.

III.   Redundancy and Ecological Plasticity

Some species, perhaps more frequently of plants than of animals, play pivotal roles in the function of particular ecosystems, and are recognized as "keystone" species. Conversely, it is common for several species, not necessarily closely related taxonomically, to contribute to a particular ecosystem process, so that an individual species in such a species association might be regarded as redundant. It is important to acknowledge that redundancy in this sense is a concept that may be limited by an anthropocentric view of a species' function or "usefulness" in an ecosystem. A species that may appear to be redundant in an ecosystem when viewed in relation to some defined parameters may have unique functional significance if related to other parameters that are not specified.

Evidence of redundancy might, nevertheless, be inferred in some earthworm species associations. In European grasslands there may be up to 15 lumbricid species (Baker 1983), each with a clearly defined niche, sharing and together utilizing the available spatial and trophic resources. In New Zealand and Australia, and perhaps in North America, where only a limited subset of the

European species has been introduced, grassland soils commonly have only one or two lumbricid species, which may attain population densities far in excess of those common for associations of numerous species in Europe. Individual species, whose niche may be clearly defined in multiple-species associations, have expanded their niches to include part or all of the "typical" niches of other species in multi-species associations. For example, Stockdill (1966, 1982) introduced *Aporrectodea caliginosa* to the soils of pastures that lacked earthworms and had probably not had significant earthworm populations since they had been cleared of native vegetation about six years previously. *A. caliginosa* and/or the closely related *A. trapezoides* are the most widespread and numerous species in New Zealand pastures (Lee 1991); they are frequently found as the only earthworm species in the soil, or in associations of two or three species. They feed on dead roots, as they do in Europe, where they deposit their casts below the surface and would be classed as endogeic species in the scheme of Bouché (1977). In New Zealand they commonly feed also on dead plant remains and the dung of herbivores, which they collect from the soil surface, depositing their casts on and below the surface. In Bouché's scheme their behavior in New Zealand would class them as part endogeic, part anecic, and part epigeic. *A. caliginosa* and *A. trapezoides* are apparently opportunistic, with the ability to expand their niche width to fill niches that might be occupied by other species if they were present. It may be concluded that the majority of species in some European species associations are redundant, since the limited number of species in the New Zealand associations apparently fill the niches occupied by the larger number of species in the European associations.

Similarly, in the sugarcane fields of northern Queensland, the introduced pantropical glossoscolecid *Pontoscolex corethrurus* (F. Muller) is usually the only earthworm species present. Green harvesting, associated with trash retention and minimum cultivation, have resulted in inputs of the order of 15 t ha^{-1}y^{-1} (dry weight) of coarse leaf and stem litter. In these circumstances large populations of *P. corethrurus* are active close to the soil surface and in the base of the litter layer; they apparently feed on partially decayed sugarcane litter and they deposit their casts above the soil surface, in or on top of the litter layer, i.e., they are epigeic/anecic species in the scheme of Bouché. In a sugar cane field near Innisfail, 81% of the litter layer (15 t ha^{-1}) had disappeared 338 days after harvesting, and its disappearance was largely due to its burial by the casts of *P. corethrurus*, with further decomposition mainly due to microorganisms (Spain and Lee, pers. comm.). In grasslands and pastures of western Africa and Central and South America, *P. corethrurus* is commonly recognized as an endogeic species (Lavelle et al. 1987).

Lavelle (1983) analyzed the functional plasticity of earthworm species and communities in Ivory Coast savannahs, and related it primarily to seasonal variability in rates of input and nutrient quality of plant litter. He showed changes in the structure of size classes, levels of activity and depth distribution in the soil of earthworms that relate to seasonal variability in the litter. He further related the ability of earthworms to modulate their behavior, and consequently their requirement for resources, to the generally low species-richness of earthworm communities and the apparently anomalous lack of meridional increase in species-richness.

IV. Management of Earthworm Populations

Existing populations of earthworms may be manipulated, i.e., managed so as to increase abundance or change species associations, indirectly or by direct intervention (Lavelle et al. 1989, Lee 1991).

A.   Indirect Intervention

Indirect intervention influences existing earthworm populations, making soil conditions more favorable for them by, for example, reduction or elimination of tillage and of the use of biocides, or irrigation, surface mulching, inputs of organic or inorganic fertilizers, raising the pH of acid soils by applying lime.

Differences in tillage practices are reflected in differences in the structure of communities and abundance of soil organisms, including earthworms. Conventional tillage incorporates most plant residues into a well aerated plough layer, where a high proportion is rapidly decomposed and the included plant nutrients are mineralized by aerobic organisms, mainly bacteria, with accompanying food webs that are basically dependent on the consumption of bacteria. Earthworms are not prominent in these food webs. Minimum tillage systems result in the accumulation of plant residues at and close to the soil surface. This promotes fungal rather than bacterial growth, which leads to the development of food webs based on fungi, to which earthworms are important contributors (Hendrix et al. 1986, Lee and Pankhurst 1992).

The influence of crop rotation and other farm practices on the abundance of earthworms is illustrated by data from the long term experimental plots at Broadbalk, Rothamsted. Edwards (1981) found that the soil under wheat, cropped continuously for 136 years, had more earthworms than soils under any of the crop rotations that had received the same input of fertilizers over the same period. A wheat crop following potatoes and beans had about half the population of the continuous wheat treatment, while in an annual fallow following wheat no earthworms were found. The Rothamsted data also show that regular use of acidic fertilizers, particularly ammonium sulphate, reduces soil pH sufficiently to depress earthworm abundance (Edwards 1981). Conversely, the application of lime to acidic soils usually promotes increase in abundance. Various investigations have shown that, among commonly used biocides, carbamates, some organophosphates, some organochlorine compounds, and soil fumigants are particularly toxic, while herbicides in general appear not to be toxic at normal rates of application. Edwards (1992) points out that results of investigations of toxicity are often not easily compared, because of differences in the methods used; the tests may be adequate to identify chemicals that are extremely toxic, but do not accurately identify moderately or slightly toxic compounds.

The Rothamsted experiments demonstrate increased earthworm abundance when animal dung is used as a fertilizer (Edwards 1981). Similarly, increased populations generally follow from stimulation of plant growth by inorganic fertilizers. Tisdall (1978) examined the effects of irrigation and input of organic matter on earthworm abundance in orchard soils in Victoria, Australia. Straw was mixed into the soil at 68 t ha[-1] and thereafter a mulch of straw and sheep dung was applied annually at the rate of 5.5 t ha[-1]; the soil was irrigated to maintain soil moisture at <30 kPa suction. After three years earthworm abundance, water infiltration rate and soil macro-porosity were all much increased compared with those at 158 sites in untreated orchards (Table 3).

B.   Direct Intervention

Direct intervention involves the introduction from elsewhere of species or strains of species that are not already present and are known to enhance soil fertility and plant growth.

Introduction of selected species of earthworms has been attempted in many countries; the results reported vary greatly and the statistical significance of much of the published data is uncertain (Brun et al. 1987). Nearly all the data listed by Brun et al. were of lumbricid species, introduced into pastures. In three cases earthworms were introduced into forest soils, but there is no record of the persistence of the earthworms, or of their long term effects on productivity.

The most comprehensive and successful introduction programs have been in New Zealand, where large scale introductions date from about 1940, based initially on the recognition by a farmer of the capability to enhance soil fertility of a particular species (*Aporrectodea caliginosa*) that was spreading across his pastures from his orchard. Long term field experiments, continued over a period of about 30 years after introduction of *A. caliginosa* (Stockdill 1966, 1982),

Table 3. Effects on abundance of lumbricid earthworms, water infiltration rate, and soil macro-porosity after three years of a combination of irrigation and mulching in orchards in Victoria, Australia (Data from Tisdall 1978)

	Earthworm Abundance No. m[-2]	Infiltration Rate Min. 50 mm[a]	Macroporosity %[b]
Irrigated and mulched orchard	2000	1	19
Control (means for 158 untreated orchards)	150	83	5

[a] Minutes for infiltration of 50 mm water from ring infiltrometer.
[b] % air-filled porosity at 4 kPa suction.

Table 4. Some effects of introduction of *Aporrectodea caliginosa* on soil properties in a silt loam under pasture in Otago, New Zealand (Data from Stockdill 1982)

Soil Property	A. caliginosa Absent	A. caliginosa Present
Field capacity (% dry weight)	42.0	51.7
Wilting point (% dry weight)	15.6	16.2
Available moisture (mm)	18.3	31.0
Infiltration rate (mm h^{-1})	14.0	26.4
Organic carbon (%)	3.54	4.46

have demonstrated sustained improvements in the physical properties of the soil (Table 4) and sustained increases of about 25–30% (Lacy 1977) in pasture production. The success of experimental introductions led to the development in New Zealand of machinery to harvest turfs from pastures with high abundance of earthworms that are active close to the soil surface (e.g., *A. caliginosa*) and to distribute the turfs on the surface of earthworm-deficient farmlands.

The New Zealand soils into which *A. caliginosa* was introduced in these experiments had been in pasture for six years or more before the introductions were made. Native megascolecid species were probably present in upper soil horizons before the pastures were established, but if so they had died out. A deep-burrowing endogeic species, *Octochaetus multiporus*, was present (Stockdill 1959). A dense superficial mat of dead roots had developed under the pasture; such root mats tend to become water repellent and acidic. *A. caliginosa* fed on the dead roots, and within about five years the root mat and surface accumulations of sheep dung had disappeared, with organic matter rather evenly distributed and soil structure upgraded, especially in the 0–10 cm soil layer.

There may be other regions where earthworms are for various reasons entirely lacking, especially perhaps in the north of North America, where the land was covered until only 12–15,000 years ago by ice sheets. It is likely, however, that species selected for introduction will more commonly be introduced into soils that already have other species of earthworms.

From the results of many investigations Brun et al. (1987) concluded that beneficial effects on soils are primarily due to anecic and endogeic species. *Aporrectodea longa* (Ude), a comparatively deep burrowing lumbricid which is distributed sporadically in New Zealand, has been introduced in field experiments (Syers and Springett 1983, Springett 1985) in a pasture with a resident population of *A. caliginosa*, *A. trapezoides*, and *Lumbricus rubellus* (Hoffmeister), which are active close to the soil surface. The trials of Springett (1985), which included the addition of lime to the soil surface at the time of introduction of *A. longa* (150 m^{-2}), resulted in large and significant increases in pasture production with lime alone, and further small significant increases when *A. longa* was introduced together with addition of lime. Root biomass at 15–20 cm depth in the presence of *A. longa* was about twice that in plots without *A. longa* (p <0.01). The increase in root biomass at depth, and the increase in macro-porosity that it implies, may be particularly important in regions where plant growth is restricted by seasonal moisture stress. Specimens of *A. longa* cannot be collected with the shallow-digging machinery developed for *A. caliginosa*, and for Springett's experiments the worms were collected from the field by hand. No suitable method for mass rearing of *A. longa* was available at the time, but a recently developed method for mass rearing of the anecic *Lumbricus terrestris* L. in plastic bags (Butt et al. 1992) might profitably be further developed and applied to *A. longa* and other deep-burrowing species.

V.   Current Research Problems

A.   Basic Research

1. Taxonomy and Biogeography: There is an urgent need for taxonomic and biogeographic studies in most parts of the world, including the United States. It is necessary to know what earthworms we have, and where they are. It is not easy to find financial support for extensive survey and collecting, which are seen by many scientists as "natural history" pursuits that demand long term support and have little scientific merit. A partial solution to this problem has been found in Australia, with collaboration between CSIRO scientists and the national school students' science club, Double Helix (Anon. 1993, Baker in press, Baker pers. comm.) in a project "Earthworms Downunder" that was conducted over several months in 1992. Collection instructions, a simple kit for postage of specimens and a simple key (Baker and Kilpin 1992) for identifying 16 species known to be common in Australia were provided and more than 1400 students cooperated. The project was given wide media exposure, and it was concluded (Anon. 1993) that the work done represented at least five years of research time. The project had the additional benefit of providing wide and favorable publicity on the practical importance of earthworms and the need for research.

2. Functional Groups: The concept of functional groups has proved useful in interpreting relationships between earthworms and between earthworms and soils. To some extent, recognition of functional groups is useful where species identification is not possible; species can be grouped on the basis of ecological and behavioral characters. If functional groups are to be compared and used to manage biodiversity on a broad geographical scale, accurate identification of the species that make up the groups is necessary. The concepts of demographic strategies, developed by Lavelle (1979), would seem to be capable of providing a key to the functional significance of species and species associations; they deserve investigation in a wider biogeographical context.

3. Intraspecific Genetic Diversity: The recognition of intraspecific genetic diversity and its possible relationship with ecological plasticity may provide a key for the selection of strains best suited for particular environments and management goals. This is a field that is not well developed, but the use of modern techniques, e.g., DNA fingerprinting, now being used for introduced lumbricids in Australia, indicate that this is a promising line of research.

4. Redundancy: Functional redundancy, i.e., coexistence of two or more species, not necessarily closely related taxonomically, whose roles in ecosystem processes have much in common, is widely recognized in the soil biota. The ubiquity of earthworms, their incorporation of plant remains into the soil, and their burrowing and casting behavior set them apart from other soil biota, in a functional ecological sense and in terms of their effects on soils and plant growth. Their effects on nutrient cycling are important, but other organisms are capable of achieving the same effects in the absence of earthworms. Within species associations that make up earthworm communities there frequently appears to be redundancy of behavior within functional groups. Ecological plasticity further complicates the possibilities for redundancy of function. There is a need for research to provide better understanding of the behavior of individual species, to determine whether apparent redundancy is real. This would have important implications for management of populations.

B.   Management of Earthworm Populations

1. Farm Management Techniques: It is widely acknowledged that reduced or zero tillage, retention of stubble or other crop wastes, the use of organic fertilizers, care in the selection and use of biocides that have minimum impact on earthworms, protection of surface soil and forest litter layers from compaction by excessive traffic and trampling, provision of shade, surface mulching, irrigation, and prevention or reduction of soil erosion all contribute to the survival and abundance of earthworms. There is a need to continue field programs that demonstrate to land users the role of these and other management techniques that favor earthworms.

Other management-related research currently in progress includes:

- better definition of the habitat requirements of species, so that management techniques can be tailored to cause minimum disturbance;
- cost/benefit assessments of the value of earthworms in terms of soil structure amelioration, nutrient availability, soil and water conservation;
- the role of earthworms in the success of alternative agricultural practices with reduced inputs of agricultural chemicals, tillage and other farm practices. Losses in productivity that may result must be offset against reduced costs of inputs and gain in sustainability of land use.

2. Introduction of Species: The history of introduction of exotic species of animals and plants to new environments is rich in examples of unexpected and

often undesirable consequences. Introduction of exotic earthworm species should not be undertaken without careful prior consideration of: (a) whether there is a problem that can be solved or ameliorated by earthworms; and (b) the possibility of undesirable consequences.

Nearly all the previous successful introduction programs have depended on collection of worms from field populations of high density. The most successful collection methods have involved harvesting and distribution of turfs from areas that contain large numbers of target species that are active close to the ground surface. These methods are not practicable for species that are active deeper in the soil, and these may often be important contributors to soil fertility and sustainable land use.

A minimum target earthworm community, including species to be introduced and species already present, might comprise:

- One or more anecic or anecic/epigeic species that make vertical or near vertical burrows that open at the soil surface and that feed and deposit their casts at the soil surface, so removing or burying plant litter.
- One or more endogeic species that feed on dead roots and other subsurface organic matter, and that make more horizontally oriented burrows.
- Among these, or in addition, at least one species that makes deep burrows and can penetrate compact soil layers.

Lee (1991) summarized the kind of information needed for the selection and successful introduction of earthworms to enhance soil fertility and productivity, as follows.

- Definition of problems that are amenable to solution through the activities of earthworms.
- Physicochemical and biological characteristics of the soil, whether these may limit earthworms, and if so whether they can be modified to counter the limitations.
- Whether species that are already established are capable of providing the benefits sought, and if so how management might be modified to stimulate the effect of these species.
- Availability and location of species or strains, not already established, that are likely to provide benefits, and are likely to thrive in the new environment.
- Understanding of the behavior of candidate species, i.e., their place in relation to "functional groups".
- Establishment of collection or culture methods capable of yielding individuals in numbers appropriate to the needs of the introduction program.

VI.   Research Imperatives

A.   Use of Earthworms as Indicators of Sustainability

The species that constitute species associations have been shown, at least in pasture soils, to vary qualitatively and quantitatively in response to changes in soil fertility and the species composition of pasture plant communities (e.g., see Lee 1959). It seems likely that they could be useful as indicators of levels of stress, and so of the sustainability of land use systems (Lee 1992).

B.   Taxonomic, Biogeographic, Ecological, and Behavioral Studies

Present knowledge of earthworm biology largely concerns a small number of common species. There are known to be about 3500 species (Reynolds and Cook 1993); if maximum use is to be made of the potential of earthworms to contribute to sustainable land use practice, there is a need for better understanding of species selected on a "best guess" basis as likely to be appropriate.

C.   Earthworm Physiology

Basic knowledge of many aspects of earthworm physiology was comprehensively reviewed by Mill (1978). The digestive capabilities and symbiotic relationships with intestinal microorganisms have received some attention (see, for example, Hartenstein 1981, Barois et al. 1987, Loquet and Vinceslas 1987), as have some other aspects of earthworm physiology. There is a need for a better understanding, especially of the digestive and excretory physiology of more species, and the application of the knowledge gained to understanding of soil chemistry, organic matter decomposition, and plant nutrient cycling.

References

Anon. 1993. Earthworm's a winner. *The Helix* Publ. No. 30, 2 pp. Double Helix Science Club, CSIRO Canberra, Australia.

Baker, G.H. 1983. Distribution, abundance and species associations of earthworms (Lumbricidae) in a reclaimed peat soil in Ireland. *Holarct. Ecol.* 6:74–80.

Baker, G.H. (in press). Introduction of earthworms to agricultural soils in southern Australia. *Proc. National Workshop on Role of Earthworms in Agriculture and Land Management, Launceston, Tasmania, 1993.*

Baker, G.H. and G.P. Kilpin. 1992. *Earthworm Identifier* 12 pp. Double Helix Science Club, CSIRO. Canberra, Australia.

Barois, I., B. Verdier, P. Kaiser, A. Mariotti, P. Rangel, and P. Lavelle. 1987. Influence of the tropical earthworm *Pontoscolex corethrurus* (Glossoscolecidae) on the fixation and mineralisation of nitrogen. p. 151-158. In: Bonvicini Pagliai, A.M. and P. Omodeo (eds.) *On earthworms*. Selected Symposia and Monographs U.Z.I. 2. Mucchi, Modena, Italy.

Bouché, M.B. 1972. *Lombriciens de France*. INRA Publ.72-2. INRA, Paris.

Bouché, M.B. 1977. Stratégies lombriciennes. In: Lohm, U. and T. Persson (eds.) *Soil organisms as components of ecosystems. Ecological Bulletin (Stockholm)* 25:122-132.

Brun, J.J., D. Cluzeau, P. Trehen, and M.B. Bouché. 1987. Biostimulation: perspectives et limites de l'amélioration biologique des sols par stimulation ou introduction d'espèces lombriciennes. *Rev. Ecol. Biol. Sol* 24:685-701.

Butt, K.R., J. Frederickson, and R.M. Morris. 1992. The intensive production of *Lumbricus terrestris* L. for soil amelioration. *Soil Biol. Biochem.* 24:1321-1325.

Calow, P. 1977. Ecology, evolution and genetics: a study in metabolic adaptation. *Adv. Ecol. Res.* 10:1-62.

Darwin, C.R. 1837. On the formation of mould. *Proc. Geol. Soc.* 5:505-509.

Darwin, C.R. 1881. *The formation of vegetable mould through the action of worms with observations on their habits*. Murray, London.

Edwards, C.A. 1981. Earthworms, soil fertility and plant growth. p. 61-85. In: Appelhof, M. (ed.) *Proceedings of a workshop on the role of earthworms in the stabilisation of organic residues*. Beachleaf Press, Kalamazoo, MI.

Edwards, C.A. 1992. Testing the effects of chemicals on earthworms: the advantages and limitations of field tests. p. 75-84. In: Greig-Smith, P.W., H. Becker, P.J. Edwards and F. Heimbach (eds.) *Ecotoxicology of earthworms*. Intercept Press, Andover, England.

Gates, G.E. 1970. Miscellanea Megadrilogica. VIII. *Megadrilogica* 1(2):1-14.

Gates, G.E. 1976. More on oligochaete distribution in North America. *Megadrilogica* 2:1-8.

Graff, O. 1953. Die Regenwurmer Deutschlands. *Schrift. Forsch. Land. Braunschweig-Volkenrode* 7.

Hartenstein, R. 1981. Production of earthworms as a potentially economical source of protein. *Biotech. Bioeng.* 23:1797-1811.

Hendrix, P.F., R.W. Parmelee, D.A. Crossley, Jr., D.C. Coleman, E.P. Odum, and P.M. Groffman. 1986. Detritus food webs in conventional and no-tillage agroecosystems. *BioScience* 26:374-380.

Jaenike, J., S. Ausabel, and D.A. Grimaldi. 1982. On the evolution of clonal diversity in parthenogenetic earthworms. *Pedobiologia* 23:304-310.

James, S.W. and M.R. Cunningham. 1989. Feeding ecology of some earthworms in Kansas tallgrass prairie. *Am. Midl. Nat.* 121:78-83.

Lacy, H. 1977. Putting new life in wormless soil. *N.Z. Farmer* 98:20-22.

Lavelle, P. 1979. Relations entre types écologiques et profils démographiques chez les vers de terre de la savane de Lamto (Côte d'Ivoire). *Rev. Ecol. Biol. Sol* 16:85-101.

Lavelle, P. 1983. The structure of earthworm communities. p. 449-466. In: Satchell, J.E. (ed.) *Earthworm ecology from Darwin to vermiculture*. Chapman and Hall, London.

Lavelle, P., I. Barois, I. Cruz, C. Fragoso, A. Hernandez, A. Pineda, and P. Rangel. 1987. Adaptive strategies of *Pontoscolex corethrurus* (Glossoscolecidae, Oligochaeta), a peregrine geophagous earthworm in the humid tropics. *Biol. Fertil. Soils* 5:188-194.

Lavelle, P., I. Barois, A. Martin, Z. Zaidi, and R. Schaefer. 1989. Management of earthworm populations in agro-ecosystems: a possible way to maintain soil quality?

p.109–122. In: Clarholm, M. and L. Bergström (eds.) *Ecology of arable land*. Kluwer Academic Publishers, Stockholm.

Lee, K.E. 1959. The earthworm fauna of New Zealand. *N.Z. Dept. Sci. Ind. Res. Bull.* 130:1–486. Government Printer, Wellington, New Zealand.

Lee, K.E. 1983. Soil animals and pedological processes. p. 629–644. In: Division of Soils, CSIRO. *Soils: an Australian viewpoint*. CSIRO, Melbourne/Academic Press, London.

Lee, K.E. 1985. *Earthworms: their ecology and relationships with soils and land use*. Academic Press, Sydney, Australia.

Lee, K.E. 1987. Peregrine species of earthworms. p. 315–327. In: Bonvicini Pagliai, A.M. and P. Omodeo (eds.) *On earthworms*. Selected Symposia and Monographs U.Z.I., 2. Mucchi, Modena, Italy.

Lee, K.E. 1991. The diversity of soil organisms. p. 73–87. In: Hawksworth, D. L. (ed.) *The biodiversity of microorganisms and invertebrates: its role in sustainable agriculture*. CAB International, Wallingford, England.

Lee, K.E. 1992. Where does biodiversity fit? p. 59–63. In: Hamblin, A. (ed.) *Environmental indicators for sustainable agriculture*. Report on a National Workshop, Nov. 28–29, 1991. Bureau of Rural Resources, Canberra, Australia.

Lee, K.E. and C.E. Pankhurst. 1992. Soil organisms and sustainable productivity. *Aust. J. Soil Res.* 30:855–892.

Loquet, M. and M. Vinceslas. 1987. Cellulolyse et lignolyse liées au tube digestif d'*Eisenia fetida andrei* Bouché. *Rev. Ecol. Biol. Sol* 24:559–571.

MacArthur, R.H. and E.O. Wilson. 1967. Some generalized theorems of natural selection. *Proc. Nat. Acad. Sci. U.S.A.* 48:1893–1897.

Mill, P.J. (ed.) 1978. *Physiology of Annelids*. Academic Press, London.

Németh, A. and R. Herrera. 1982. Earthworm populations in a Venezuelan tropical rain forest. *Pedobiologia* 23:437–443.

Nordström, S. and S. Rundgren. 1974. Environmental factors and lumbricid associations in southern Sweden. *Pedobiologia* 14:1–27.

Odum, E.P. and L.J. Biever. 1984. Resource quality, mutualism and energy partitioning in food chains. *Am. Nat.* 124:360–376.

Omodeo, P. 1952. Cariologia dei Lumbricidae. *Caryologia* 4:173–275.

Omodeo, P. 1953. Cariologia dei Lumbricidae. II. *Caryologia* 8:135–178.

Phillipson, J., R. Abel, J. Steel, and S.R.J. Woodell. 1976. Earthworms and the factors governing their distribution in an English beechwood. *Pedobiologia*, 16:258–285.

Pianka, E.R. 1970. On *r*- and *K*-selection. *Am. Nat.* 104:592–597.

Reinecke, A. J. 1975. The influence of acclimation and soil moisture on the temperature preference of *Eisenia rosea* (Lumbricidae). p. 341–349. In: Vanek, J. (ed.) *Progress in soil zoology*. Junk, The Hague.

Reynolds, J.W. and D.G. Cook. 1993. *Nomenclatura Oligochaetologica. Supplementum Tertium*. vi+33 pp. New Brunswick Museum Monographic Series (Natural Science) No. 9.

Springett, J.A. 1985. Effect of introducing *Allolobophora longa* Ude on root distribution and some soil properties in New Zealand pastures. p. 399–405. In: Fitter, A.H., D. Atkinson, D.J. Read, and M.B. Usher (eds.) *Ecological interactions in soil*. Blackwell, Oxford.

Stockdill, S.M.J. 1959. Earthworms improve pasture growth. *N.Z. J. Agric.* 98:227–233.

Stockdill, S.M.J. 1966. The effects of earthworms on pastures. *Proc. N.Z. Ecol. Soc.* 13:68–75.

Stockdill, S.M.J. 1982. Effects of introduced earthworms on the productivity of New Zealand pastures. *Pedobiologia*, 24:29–35.

Syers, J.K. and J.A. Springett. 1983. Earthworm ecology in grassland soils. p. 67–83. In: Satchell, J.E. (ed.) *Earthworm ecology*. Chapman and Hall, London.

Tisdall, J.M. 1978. Ecology of earthworms in irrigated orchards. p. 297–303. In: Emerson, W.W., R.D. Bond, and A.R. Dexter (eds.) *Modification of soil structure*. Wiley, Chichester, England.

van Breemen, N. 1993. Soils as biotic constructs favouring net primary productivity. *Geoderma* 57:183–211.

Systematic Index

Subject Index